# ANIMAL MOTIVATION

# Chapman and Hall Animal Behaviour Series

SERIES EDITORS

**D.M. Broom**

*Colleen Macleod Professor of Animal Welfare, University of Cambridge, UK*

**P.W. Colgan**

*Professor of Biology and Psychology, Queen's University at Kingston, Canada*

Detailed studies of behaviour are important in many areas of physiology, psychology, zoology and agriculture. Each volume in this series will provide a concise and readable account of a topic of fundamental importance and current interest in animal behaviour, at a level appropriate for senior undergraduates and research workers.

Many facets of the study of animal behaviour will be explored and the topics included will reflect the broad scope of the subject. The major areas to be covered will range from behavioural ecology and sociobiology to general behavioural mechanisms and physiological psychology. Each volume will provide a rigorous and balanced view of the subject although authors will be given the freedom to develop material in their own way.

# ANIMAL MOTIVATION

**Patrick Colgan**

*Professor of Biology and Psychology,*
*Queen's University at Kingston, Canada*

London   New York
CHAPMAN AND HALL

First published in 1989 by Chapman and Hall Ltd
11 New Fetter Lane, London EC4P 4EE
Published in the USA by Chapman and Hall
29 West 35th Street, New York NY 10001

© 1989 Patrick W. Colgan

Set in 11/12 pt Bembo by Best-Set, Hong Kong
Printed in Great Britain at the University Press, Cambridge

ISBN   0 412 31850 4 (Hb)   0 412 31860 1 (Pb)

British Library Cataloguing in Publication Data

Colgan, Patrick
Animal motivation. – (Chapman and Hall
animal behaviour series).
1. Animals. Motivation
I. Title
156'.38

ISBN 0-412-31850-4
ISBN 0-412-31860-1 Pbk

Library of Congress Cataloging-in-Publication Data

Colgan, Patrick W.
Animal motivation/Patrick Colgan.
p.    cm. – (Chapman and Hall animal behaviour
series)
Bibliography: p.
Includes index.
ISBN 0-412-31850-4.   ISBN 0-412-31860-1 (pbk.)
1. Motivation in animals.   I. Title.   II. Series.
QL781.3.C65   1989
591.5'1—dc19

# Contents

*Preface*                                                                  vii
*Acknowledgements*                                                          ix

**Introduction**                                                            1
1.1   Tinbergen's framework                                                 1
1.2   The key processes                                                     3
1.3   Historical background of ethology and psychology                      4
1.4   The concepts of instinct and drive                                    6

**Ethology of motivation**                                                 11
2.1   Some behavioural concepts                                            11
2.2   Ontogeny of feeding in chicks                                        25
2.3   Motivation as a constraint on learning                               32
2.4   Combination of factors                                               37
2.5   Interactions between motivational systems                            46
2.6   Quantitative models for motivational processes                       56
2.7   Cognitive approaches                                                 63

**Physiology of motivation**                                               69
3.1   Invertebrate systems                                                 70
3.2   Homeostasis                                                          77
3.3   Arousal and the reticular activating system in cats                  80
3.4   Hunger and brain lesions in rats                                     86
3.5   Hormones and sexual drives in canaries and lizards                   92
3.6   Conclusions                                                          98

**Ecology of motivation**                                                 101
4.1   Optimal behaviour                                                    102
4.2   Matching and maximizing                                              105
4.3   Individual differences                                               115
4.4   Hunger and foraging                                                  118
4.5   Signalling of intentions                                             123

**Overview**                                                          133

*References*                                                        137
*Author index*                                                      153
*Subject index*                                                     157

# Preface

The topic of animal motivation deals with how and why animals engage in particular activities: what mechanisms inside the animal generate behaviour, how stimuli from the external environment influence these mechanisms, and how this behaviour is beneficial to the animal. The topic is thus central both to academic studies in psychology and zoology and to applied matters in domestic species. Motivation has not been an area of great emphasis in the past 10–15 years but there is now a growing realization that it should receive greater attention. Drawing on concepts and observations from a number of areas, this book provides an overview of the motivational processes which determine the choices, timing, and sequencing which are characteristic of animal behaviour. Data and theory from ethology, psychology, and evolutionary biology are synthesized into a contemporary framework for analysing such central features of behaviour as persistence in activities and goal orientation. Principles of motivational analysis are discussed and illustrated with specific case studies. The successive chapters deal with ethological, physiological, and ecological approaches involving experimental work on a diversity of vertebrate and invertebrate species. Ethological topics include the interaction of external stimuli and internal states, mechanisms of choice, quantitative models of motivation, conflict between tendencies for different activities, and behavioural homeostasis. The review of physiological research focuses on hunger, the activating roles of nerves and hormones, and the examination of animals with small nervous systems. The ecology of motivation deals with such problems as the optimization of behaviour, especially foraging, individual differences in behaviour, and the motivation for social communication. Controversies surrounding cognitive ethology and concepts such as drive and instinct are also considered. Through this coverage are presented the dimensions of motivated behaviour, the proximate physiological mechanisms for this behaviour, and the functional importance of such behaviour to an animal in its natural setting. These aspects are illustrated with case studies, and likely avenues for future developments are outlined. The book

is intended to be broad yet succinct, without trying to be encyc-
lopaedic, for upper-year undergraduates, postgraduate students, and
teachers and researchers in animal psychology and behavioural
biology.

# Acknowledgements

I am grateful to S. Crawford, R. Eikelboom, L. Gabora, C. Harris, J. Hogan, J. Morgan, H. Nogrady, L. Ratcliffe, M. Stayer, J. Sutcliffe, and J. Templeton for assistance with the literature and comments on the text. S. Crawford, J. Frame and L. Jamieson prepared the manuscript with skill and patience. Financial support was provided by an Operating Grant from the Natural Sciences and Engineering Research Council of Canada. Family and friends, especially Andrew and Jeffrey, were supportive in much-appreciated ways.

# 1
# *Introduction*

In the lengthening days of spring, a male canary sings its characteristic tunes. In a nest excavated from the gravel, a pair of cichlid fish
engage in elaborate rituals of courtship. In a quiescent group of
rhesus monkeys, an infant leaves its mother, explores a stone, and
then returns and suckles. In each of these and a multitude of other
fascinating instances, animals display complex abilities to switch
from one activity to another, to modulate the intensity and patterning of their responses, and to direct their activity toward specific
goals. Together, these abilities reflect a collection of processes glo-
bally referred to as motivation. The purpose of this book is to
examine this central topic in animal behaviour.

Let us begin with the setting within which we shall explore the
topic of motivation. The spectrum of contemporary biology is a
broad one, extending over all the levels of organization from bio-
chemistry and physiology through ethology to population and community ecology. The study of animal behaviour occupies a position
between whole-animal physiology and behavioural ecology. This
introduction serves to present the viewpoint from which motivation
will be examined in this book. The viewpoint can be appreciated
by considering Tinbergen's framework; definitions of the key processes; the historical background of ethology and psychology; and
the concepts of instinct and drive.

## 1.1 TINBERGEN'S FRAMEWORK

Much of the work in ethology has been carried out within the
framework of Niko Tinbergen's (1963) four focal points of causation, survival value, ontogeny, and evolution. For instance, when
considering squirrels feeding on acorns, these four points raise the
following questions:

1. how do stimulus cues from the acorn together with hunger cues
   from inside the squirrel cause feeding?
2. how does the pattern of feeding enhance the survival of the
   squirrel?

3. how does the pattern develop over the lifetime of the squirrel?
4. how have feeding patterns evolved in rodents generally?

Within this framework, many topics in animal behaviour can be distinguished, and are dealt with in the volumes of this series. As we shall see, motivation is a topic belonging to the area of causation, and deals with the initiation, persistence, and goal-directedness of behaviour. In quantitative terms, motivational aspects of activities include their latency, duration, intensity, and frequency.

At each level of organization, biology begins with descriptions of phenomena under study, and then proceeds to consider the causation of these phenomena. Ethology is a young science and deals with complex phenomena, so it is scarcely surprising that much description remains to be carried out. In terms of the study of causation, it is important to note that biology in general, and Tinbergen's framework in particular, have two major facets, i.e. **how** and **why** questions. These reflect the pursuit of **proximate and ultimate causation**, the mechanisms underlying biological events and their functional and evolutionary origins. Consider as an example the study of bird song. A mechanistically oriented biologist will provide an account in terms of hormonal and neural mechanisms (how) whereas one who is concerned with function and evolution will discuss the role of song in territoriality and attracting mates (why). The major contribution of Konrad Lorenz and Niko Tinbergen was in providing a synthesis of these two facets in what has become known as **classical ethology**. We shall examine later the details of their synthesis which has been elaborated by many contributors, particularly Robert Hinde.

In terms of ultimate causation, it is plausible to assume that natural selection of traits contributing to biological fitness has produced within an animal a set of interacting causal systems which function to deal with the major problems encountered. These can be grouped into four areas:

1. nutrition and fluid balance (food and water);
2. reproduction (courtship and parental care);
3. aggression (fighting and fleeing); and
4. personal care (sleep, temperature selection, and care of the body surface).

Subsequent sections of this book will focus on various aspects of these systems.

## 1.2 THE KEY PROCESSES

The key processes upon which we shall focus are stimuli, motivation, and behavioural patterning. **Stimuli** are proximate external or internal events which produce changes in overt behavioural activities. The causal states generated by these stimuli are collectively termed **motivation**. When an animal responds differently to the same stimulus on different occasions, the underlying process may be a consequence of either a structural change, such as maturation or learning, or a motivational change, i.e. a nonstructural change due to variation in drive or arousal level as reflected in the ease with which responses can be elicited or sustained. Thus there is a distinction between associative changes in mechanisms and non-associative changes in the activation of these mechanisms. Associative changes refer to alterations in the structural linkages which connect stimuli and responses while non-associative changes produce the patterning of behavioural dynamics. Additionally, the time scale of structural changes tends to be longer than that of motivational changes. Separating structure and motivation is not always easy since we are largely ignorant of the underlying physiological processes and must infer the existence of such internal characters as intervening variables from observable, external input and output measures (MacCorquodale and Meehl, 1948, discussed further at the beginning of Chapter 2). The intended purpose of using such variables is to help account for the diversity of observed behaviour with a few convenient central concepts that summarize the available data. Thus, for example, the motivation to feed or to fight can be modelled as the output of an explicitly described system. The development of such models is discussed in section 2.6.

**Behavioural patterning** refers to the relations among the latencies intensities, frequencies, durations, and sequences of acts. Students of motivation are interested in the influence of internal and external factors on this patterning. This interest runs at three chief levels of organization. The first deals with the physiological mechanisms responsible for sensitivity to the environment, both internal and external to the animal, and the production of the behavioural output. The second level focuses on the behavioural patterning itself, through analysis of its measurable temporal and sequential features. The third level considers the functional aspects of motivation, i.e. how behavioural activity adapts the animal to cope with the exigencies of survival and reproduction. Each of these levels has its own integrity and will be considered in a subsequent chapter. Thus, this book is organized around major problems and approaches encountered at each of these three levels of investigation, with taxonomic examples and historical, including biographical, information incorporated to illustrate these issues.

## 1.3 HISTORICAL BACKGROUND OF ETHOLOGY AND PSYCHOLOGY

This presentation of the intellectual setting for motivational analysis is complemented by a survey of its historical background. Interest in motivation is coeval with human curiosity. Concepts such as instincts, drives and urges have long been invoked to explain behaviour, and are discussed by early authors such as Aristotle and Pliny. Among Charles Darwin's many contributions to science, Chapter 6 of *The Origin of Species* is entitled 'Instincts', reflecting the importance of observations on behaviour for his evolutionary arguments. His chief behavioural work, the 1872 *Expression of the Emotions in Man and Animals*, in which Darwin detailed ethological continuity across species, included much discussion of motivation. Across the Atlantic Ocean, Wallace Craig (1876–1954) was an American behavioural biologist active in the early decades of this century in the study of bird behaviour. He made the key distinction between appetitive stimuli, which produce searching behaviour for particular goals, and consummatory stimuli, which bring behavioural sequences to a close. Building on the contributions of Craig and his European ornithological counterparts, Oskar Heinroth (1864–1944) and Julian Huxley (1887–1975), Konrad Lorenz and Niko Tinbergen articulated a viewpoint for a complete ethology based on the concepts of instinct, fixed action pattern, and releaser, each discussed below.

Ethology has sometimes been regarded as unsystematic, lacking in theoretical rigour. This charge has been made even in sober, well considered critical evaluations such as that by Philip Kitcher (1985). In fact, not only have ethologists, especially Tinbergen with his four focal points, generated a logical framework and programme of research, but this system has even been found to be useful outside the discipline. For instance, James Wilson and Richard Herrnstein, political scientist and operant psychologist respectively, have used the four aims in order to organize their detailed examination of human criminality (1985).

The study of motivation has been pursued not only by ethologists working within a zoological framework, but also by psychologists. Ethologists have tended to conduct investigations of natural behaviour in the field while psychologists have usually examined animals under more constrained situations in the laboratory. Psychology has grown out of philosophy, and Immanuel Kant (1724–1804), the foremost philosopher of the German Enlightenment, recognized three fundamental psychological faculties: conation (striving or will-ing), cognition (knowing), and affection (emotion). Motivation was central to the first faculty and, often linked with the topic of emo-tion, figured largely in the theories of early animal psychologists. Prominent among these was William McDougall (1871–1938), whose notion of instinct included all three faculties and who em-phasized the purposiveness of behaviour. An Englishman teaching German psychology in America, McDougall (1923) insisted that 'the healthy animal is up and doing', rather than a sitting sack of reflexes instigated to action only by external stimuli. His list of the seven chief characteristics of behaviour included such motivational features as spontaneity, persistence, variation, and goal-directedness and achievement. Upon these features he based his **hormic theory** of purposive behaviour, etymologically derived from a Greek term for 'urge to action'. The purposiveness of behaviour was also a dominant theme in the writings of E.C. Tolman (1886–1959), a prominent researcher into animal learning (e.g. Tolman, 1951).

Associationist philosophy is a dominant lineage stemming from John Locke (1632–1704), who maintained that mental activity de-pended on the association of ideas and impressions. This philosophy has had a major influence in the study of animal learning, especially as advanced by the American pioneer E.L. Thorndike (1874–1949) who stressed the formation of associations between stimuli and re-sponses (e.g. Thorndike, 1911). The learning theories of Clark Hull (1884–1952) and his intellectual heirs centred on two concepts, habit strength, reflecting associations; and drive strength, reflecting mo-tivation (Hull, 1952). By way of contrast, B.F. Skinner (e.g. 1974) and his disciples have eschewed theories of learning and motivation, and concentrated on the control of performance through schedules of reinforcement. It is easy to understand Skinner's abhorrence of the metaphysical quagmires in which the analysis of behaviour has frequently become bogged down. Nonetheless, most investigators regard the totally atheoretical, phenomenological approach of Skin-ner, with its restriction to 'just the data', as insufficient to lead to a full comprehension of motivation.

Major psychological texts on motivation since 1945 include Bindra (1959), Brown (1961), Young (1961), Atkinson (1964), Cofer and Appley (1964), Atkinson and Birch (1970), Bolles (1975), and Toates (1986). In the study of how individuals learn, psychologists have paid much attention to the topic of how motivation can be acquired through experience. For instance, Richard Solomon (1982) has de-veloped a theory of opponent processes, in which opposing hedonic tendencies control the affect, or emotional reaction, during and after stimulation, and thereby mediate reinforcement and incentive properties of stimuli. Finally, within applied psychology, much motivational research has investigated various ways to make employees work harder, students learn more, and consumers buy more. While some general principles such as cost–benefit analysis may be useful in both ethology and applied research, the details of this work are relevant to students of animal behaviour only at the exceptional anthropomorphic risk of endowing animals with the desires of contemporary humans.

Since 1945, a number of intellectual developments have had an impact on the study of motivation. Four of these advances are particularly important to identify:

1. approaches to the nervous and hormonal systems of animals;
2. quantitative methods originating with systems theory and cybernetics;
3. the growth of sociobiology; and
4. cognitive ethology.

Each of these developments will be discussed in the chapters which follow. In each chapter, a number of case studies involving different species will be presented from the recent literature. These studies have been selected from the mass of available data on causal systems (such as feeding, drinking, courtship, aggression, defence, and exploration) and associated emotions (hunger, thirst, lust, anger, fear, and curiosity) in many species. They serve, like Beatrice in Dante's Inferno, as a guide to illustrate the principles and important topics within motivational research.

## 1.4 THE CONCEPTS OF INSTINCT AND DRIVE

The terminological and conceptual difficulties of motivational research have been legion. Continuing disagreement over which topics properly fall within this field of research, and what is known with reasonable certainty about each, are symptomatic of these difficulties. This murkiness makes the study of motivation baffling to the many and challenging to the (fool) hardy few. Every scientific discipline, as it attempts to carve nature better to the joint, goes through the process of deriving a valid, succinct, and comprehensive vocabulary from the everyday language of its provenance. In the study of motivation, this process is, not surprisingly, proving to be a long and painful task. Reams have been written attacking and justifying various usages of motivational terms. Sometimes these usages have reflected alternative and defensible theoretical stances. In other cases there can be no doubt that many terms have been what Francis Bacon labelled 'idols of the market place'; verbal tyrants whose influence is difficult to escape. While there appears to be little profit in extensive sojourns into the Byzantine intricacies of terminology, it is worthwhile to consider the key concepts of instinct and drive.

Within ethology, Tinbergen's (1951) definition of instinct remains a historically important benchmark for contemporary theorizing on the causation of natural behaviour:

> a hierarchically organized nervous mechanism which is susceptible to certain priming, releasing and directing impulses of internal as well as of external origin, and which responds to these impulses by co-ordinated movements . . .

Such a structural concept readily supplies a framework within which motivation can be studied. Like the term innate, 'instinct' has been used in many senses (Cassidy, 1979). The continuing value of Tinbergen's definition is that it is based on well established features of neural hierarchies while at the same time incorporating the causal and functional features of behavioural organization.

Notwithstanding the influence of Tinbergen's definition, many students of motivation prefer to abandon the term instinct altogether since a behavioural activity is often described as instinctive when it is not obviously affected by environmental factors, and yet all activity is the result of a genome interacting with an environment. As with problems of ethological taxonomy, discussed in Chapter 2, it is clearly important to distinguish carefully whether a term such as instinct refers to a description, a physiological mechanism, a function, a developmental process, or an attribute of a population of animals (e.g. a behavioural pattern characteristic of all normal adult individuals). Standardization of vocabulary is surely a necessary goal if ethology, like other disciplines, is to avoid becoming a Babel of Humpty Dumpties, from Alice's Wonderland, each using a different definition.

In a thoughtful essay, Epstein (1982) has proposed that both similarities and differences exist between instinct and motivation. Both include innate mechanisms and acquired components, display sequential organization of drive-induced appetitive and consummatory phases, and contribute to homeostasis. By contrast, argues Epstein, instinct is species-specific in its sensorimotor organization, taxonomically common, and nonaffective (i.e. unaccompanied by emotion), and its appetitive phase is unmodified by expectancy; whereas motivation is individuated (i.e. varies across individuals), is taxonomically uncommon, and includes the anticipation of goals and the expression of affect (emotion). It is not clear whether Epstein's distinction can be rigidly maintained or whether instinct and motivation belong on a continuum, but he has certainly highlighted major characteristics of motivated behaviour.

A consideration of instinct leads inevitably to that of reflex, as instanced by such cases as the patellar reflex. Like instinct, the term 'reflex' has been heavily employed by a diversity of researchers. This term originated in the mechanistic theory of action proposed by Rene Descartes (1596–1650) in referring to the reflection of a stimulus into a response. Descartes provided the celebrated example of a child reacting to the heat of a fire. As the central concept in associationist analysis, the reflex dominated the materialist philosophy of mind epitomized by Descartes' fellow countryman LaMettrie; the classical work of physiologists such as I.P. Pavlov (1927) and his followers in Russia and C.S. Sherrington (1947) in Britain; as well as subsequent behaviourism. The role of reflexes in animal orientation has been the focus of attention for many years, from the early work of Jacques Loeb (1918) and Fraenkel and Gunn (1961) to recent reviews by Schone (1984) and Bell (1989). In ethology the traditional distinction has been between reflexes, viewed as fairly rigid responses elicited by specific stimuli, and instincts, more complex processes motivated at higher levels by suites of internal and external factors and resulting in an array of patterned output.

In both ethology and psychology, the term 'drive' has, like instinct, also experienced such vicissitudes of fashion that the topic provides ample grist for the mills of intellectual historians. The term was introduced into psychology by Woodworth in 1918 in his influential text *Dynamic Psychology*, and this energy concept was also central to the systems of Lorenz and Tinbergen, as well as to those of Hull and Freud. Hinde (1960) argued that the concept had no role to play in motivational analysis, but McFarland (1970) saw it as a logically inevitable component of his state–space approach, and the term remains in common use. Controversy also continues about whether motivation consists of a single drive or a set of interacting but separate drives. A detailed historical review of how the concept of drive has developed in psychology has been provided by K. Smith (1984). He concludes that it is a useful term to focus on bodily states which instigate action and underlie reinforcement of learned responses. Overall, then, an assiduous use of drive seems essential in order to comprehend the diversity of motivated behaviour which animals exhibit. Models considered in Chapter 2 will clarify how drives can be explicitly incorporated into quantitative representations of motivational systems.

Here, then, are the historical setting and the key processes and concepts with which we can begin our examination of motivation.

We turn first to the features of motivated behaviour before going on to consider its physiology and ecology.

# 2

# *Ethology of motivation*

Behaviourists are interested in the analysis of observations on animal activities in terms of behavioural concepts which encompass both physiological mechanisms (Chapter 3) and functional considerations (Chapter 4) but which also stand by themselves as descriptive and explanatory ideas. While psychologists have tended to collect data from animals in experimental situations, ethologists have generally made their observations on relatively unrestrained individuals in natural surroundings. Over the past number of years, these two traditions have come much closer in their techniques and aims, to produce a richer and more fruitful science of animal behaviour. Notwithstanding the anticipations of E.O. Wilson (1975), it is clear that physiology and sociobiology will not gobble up ethology from above and below.

## .1 SOME BEHAVIOURAL CONCEPTS

In a review of the ethology of motivation, it is helpful to understand the roles of quantification and cognitivism, the procedures of ethological taxonomy, and the tools of sequential and temporal analysis.

### Quantification and cognitivism

Two recent sources for behavioural concepts have been (1) a quantitative framework and (2) cognitivism. As soon as any motivational problem is considered in detail, it becomes important to postulate relevant states and processes. To be clear and explicit about these postulates, it is necessary to make them quantitative, and to specify how these states change over time, how the processes are affected by various factors, and how they relate to each other. As has been the case for most of biology and psychology thus far in their historical development, the quantitative tools used have been chiefly imported from other disciplines. Two imports of particular note are those of systems theory, a very general scheme which views animals as systems of interacting elements (Klir 1972), and cybernetics, the science of control and communication (Calow 1976). These approaches, fathered by Ludwig von Bertalanffy (1901–1972) and Norbert Wiener (1894–1964) respectively, enable a quantification of goal-orientation, one of the chief features of motivated behaviour (e.g. Wiener, 1961; Bertalanffy, 1968). A number of early electrical models were assembled by investigators interested in control. Such models included a heliotropic machine reported in 1915 by Jacques Loeb (1918), who was famous for his mechanistic approach to behaviour; the homeostat of Ross Ashby (1952); and the *Machina speculatrix* of Grey Walter (1953). All of these models exhibited purposeful behaviour and thus showed the intellectual affinity of cybernetics and motivational analysis.

In many behavioural situations, control is achieved through feedback and feedforward mechanisms. A furnace and thermostat constitute a familiar example of a negative feedback system. An example of feedforward is the use of peripheral receptors to detect changes in some environmental feature such as temperature and so to activate regulatory mechanisms before the internal equilibrium of the animal is disturbed. These mechanisms contribute to the maintenance of equilibria in animals and hence to homeostasis (discussed in section 3.2). Such control has been well analysed in sensory, motor, and orienting systems of various species, both vertebrate and invertebrate (e.g. McFarland, 1971; Schone, 1984; Barnes and Gladdon, 1985;). It seems likely that such structural analysis can be extended to include motivational dynamics by incorporating components representing aspects of arousal. For instance, in the Barnes and Gladdon volume, Graham Hoyle has described the production of octopamine by a single dorsal unpaired median neurone in locusts. This compound potentiates muscular contractions at neuromuscular junctions, and hence behavioural activity. Hoyle has suggested that the accumulation of octopamine in specific neurones within the locust is the physiological basis by which arousal is enhanced in these insects. The clearly homeostatic motivational systems of feeding and drinking have been modelled with control theory, and attempts to encompass less obviously homeostatic systems such as mating have also been made by focusing on processes underlying hormonal levels, intromission, and ejaculation (Toates, 1980). Mating behaviour in male rats will be examined in section 2.6.

Cognitivism has risen in popularity over the past decade in both ethology and human psychology as a reaction to what has been perceived as the shortcomings of behaviourism. The details and impact of cognitivism will be considered in section 2.7.

## Ethological taxonomy

Taxonomy is fundamental to any discipline. As in all taxonomy, there are ethological splitters and lumpers. Excessive splitting can be corrected by subsequent lumping, whereas the converse is not possible. By summarizing the basic units of behaviour, the ethogram serves a function analogous to the centimetre–gram–second system in physics or the periodic table in chemistry. In testimony to the continuing usefulness of Tinbergen's definition of an instinct as a hierarchy, it is often useful to identify basic acts which combine to form actions which in turn produce activity patterns (S. Crawford, in prep.). The totality of these patterns is behaviour, a term with no plural, common misuse notwithstanding.

The establishment of a valid ethological taxonomy is a central task for behavioural analysis. It may be, as many philosophers of science maintain, that we are inevitably biased by the theoretical tenets which we hold, at least to some extent. Nevertheless, being as empirical or purely phenomenological as we can, we pursue ethological analysis by observing the activities of our study animals until we feel well acquainted with them. We then set up an ethogram, a catalogue of ethological units, of repeatedly recognizable actions with variability about some typical value. These are often called modal action patterns, or MAPs (Barlow, 1977). The MAPs are mutually exclusive and exhaustive; each act by an animal can be described as an instance of exactly one MAP.

By way of illustration of a MAP, Figure 2.1 shows the cumulative distribution of 20 instances of yawning in each of four male sticklebacks. Both the means and variances of the distributions differ over the four males: such individual variation is discussed in section 4.3.

Table 2.1 provides an instance of an ethogram. More refined ethograms can be established, like choreographic notations, by describing acts in terms of the positions and movements of body parts with respect to each other, as well as relevant aspects of the environment such as conspecifics in social interactions. Schleidt *et al.* (1984) have developed such an ethogram for the bluebreasted quail (*Coturnix chinensis*), while Yaniv and Golani (1987) have analysed aggressive interactions in honey badgers (*Mellivora capensis*) in this way. Methodological problems encountered in forming ethograms have been considered by Drummond (1981). The examination of MAPs also illustrates the links between ethology and allied disciplines. Unravelling the sequences of muscular contractions of which MAPs consist, through such techniques as electromyographic re-

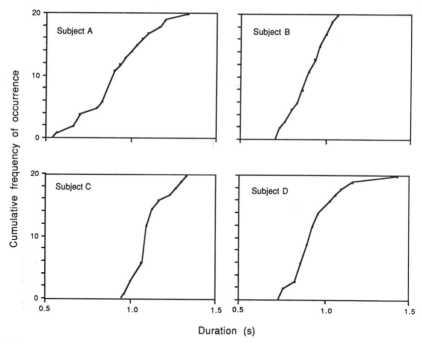

**Figure 2.1** Cumulative distributions of 20 instances of the modal action pattern, yawning, in each of four male sticklebacks (from Crawford and Colgan in preparation).

**Table 2.1** Ethogram of MAPs involved in feeding behaviour by young centrarchid fish (from Brown and Colgan, 1985).

| Feeding MAP | Description |
| --- | --- |
| Orientation | Response of fry to prey item which sometimes involves a movement of the trunk of the fry in longitudinal alignment with the prey. |
| Fixate | Pause between orientation and capturing act. |
| Lunge | Prey-capturing response in which the trunk of the fry assumes an S-shaped position and results in a rapid forward movement of the fry. |
| Snap | Prey-capture response similar to lunge except the trunk of the fry is not as tightly coiled in the S-shaped position. Fry darts forward slightly. |
| Bite | Prey-capture response which does not involve any noticeable change in forward speed. Fry opens and closes mouth quickly. |

cording, highlights the connections between ethology and functional morphology, while the study of the heritable basis of these actions leads to the field of behavioural genetics.

MAPs are the ethological descendants of FAPs, fixed action patterns, which are a major category within Lorenz's causal taxonomy of actions. Other categories are reflexes and orienting responses (for which see Bell, 1989). According to Lorenz, FAPs have several defining features. They are found in stereotyped form in all normal members of a species, are genetically determined, and develop ontogenetically without the need for any specific experience. They occur as the result of hierarchically organized instinctive drives in response to specific stimulus complexes, termed releasers. Once under way, a FAP is not influenced by environmental cues, and any variation which is observed is due to taxis, or orientation, which is not a part of the FAP itself. FAPs are consummatory responses, such that their occurrence decreases the likelihood of an early re-occurrence, while non-occurrence increases the likelihood of occurrence. The spontaneity of FAPs raises the possibility of occurrences in the complete absence of external stimuli. Such vacuum responses are discussed in section 2.4. FAPs function to enable the animal to achieve biologically important goals such as feeding, mating, rearing offspring, and displacing rivals. Finally, given that they are encoded in the genes, FAPs can be used to identify evolutionary homologies (similarities between species due to descent from a common ancestor) and so can assist systematists wishing to establish phylogenies.

The paradigmatic case of FAP and taxis, an orientation response, is egg retrieval in the greylag goose (Figure 2.2). When an egg rolls out of a nest (or is moved by a meddling ethologist) the incubating bird rises, extends its neck, and retrieves the egg by pushing it gently with the underside of its bill. Lorenz (1971) distinguished between the FAP operating in the sagittal plane and a taxis operating in the transverse plane. The FAP is the result of the nesting instinct while the taxis is an orienting response to maintain the bill beneath the egg. Thus acting as a stimulus, the egg both releases and directs responses. The separateness of these components is shown by removal of the egg once retrieval is under way. The taxis disappears but the FAP is completed, showing its instinctive basis. At the core of Lorenz' synthesis, FAPs embody structural, causal, ontogenetic, genetic, functional, and evolutionary aspects of ethology. While this synthesis provided a powerful and useful impetus for much subsequent work, it is now realized to be flawed in a number of respects, especially on genetic and ontogenetic grounds. In particular, the distinction between instinctive and learned components of behaviour, which Lorenz stressed, is not viable. Further, the development

**Figure 2.2** Egg retrieval in the greylag goose (from Lorenz, 1971).

of behavioural patterns over the lifetime of an animal, their genetic bases, and the role of stimuli in eliciting and directing them, are all more complex than he believed. Of relevance to behavioural taxonomy is the realization that FAPs are not fixed but, like all biological characters, vary within and between individuals. The concept of MAPs is intended to reflect this variation.

## Sequential analysis

Once a behavioural catalogue has been established, the analysis of the dynamics of activities can be undertaken including aspects such as the stereotypy, or fixity, of behavioural sequences. Approaches to this task vary according to the orientation of the researcher. For instance, Thompson and Lubinsky (1986) provided an interpretation of such dynamics in terms of operant conditioning. Ethological analysis has tended to take one of two routes, focusing on sequential and temporal aspects of data. The first has been to view the acts in sequences and to deal with these sequences through various devices. Richard Dawkins (1976) has argued that hierarchical organization is a major feature of ethological systems and that this realization can guide the choice of such analytical devices. A hierarchy consists of a system of entities forming sub-associations which are bound by co-ordinating links at higher and higher levels. A decorative mobile is a physical example; a phylogenetic tree is an evolutionary example; the behaviour of *Pleurobranchaea* discussed in section 3.1 is an ethological example, and governmental bureaucracies, military forces, and episcopal churches are social examples. The evidence for hierarchical organization in the nervous system was first emphasized by Paul Weiss and employed by Tinbergen in his definition of instinct. Beyond this neuronal evidence for such organization, Dawkins presents a number of theoretical arguments which lead to the expectation of hierarchies. Functionally, hierarchies and sub-assemblies permit quick and efficient local administration of responses, and decrease overall redundancy in the nervous system. From an evolutionary viewpoint, the sub-assemblies of a hierarchy may have separate genetic bases and thus can be selected relatively independently and rapidly. The issue of mechanisms for switching between hierarchically organized causal systems is discussed in section 2.5.

The use of grammars is one possible device for analysing sequences of acts. A grammar is a set of rules specifying the allowable sentences which can be generated by the elements of a language. The theory of grammars has been elaborated in the fields of linguistics and mathematics and, as Rodger and Rosebrugh (1979) have illustrated, many possible grammars can be applied to behavioural sequences. Various algorithms exist for deriving a grammar from a set of observations, and these derived grammars can indicate which production mechanisms are operative. For instance, since performance often leads to fatigue, perhaps transitions between acts which share elements are reduced by fatigue of these joint elements.

Another device for sequential analysis is the pre-post state histogram, in which plots are generated of the frequency with which one act occurs as a function of its proximity to a specified act (Douglas and Tweed, 1979). As shown in Figure 2.3, the plots present frequency against the lag both before and after the specified act. The histogram indicates at what positions in a sequence one act, in this case resting, occurs relatively often or rarely with respect to a second act, in this case feeding. Statistical tests for significant associations between acts can be carried out by setting up a Markov probability model under which the animal switches from one act to another according to a fixed probability. (Markov models will be considered in detail in section 2.6.) The histograms thus provide visual presentations of the direct associations among acts and can be linked to formal statistical models.

The relations among acts can also be investigated by examining matrices of transition frequencies between acts, or of correlations of their occurrence, with multivariate tools (such as clustering, multidimensional scaling, and factor and principal components analyses; see Colgan, 1978) in order to detect underlying causal factors. Ethol-

Ethology of motivation

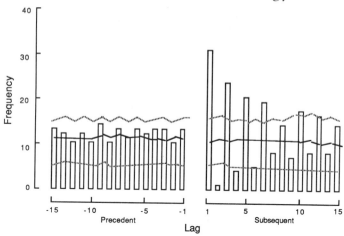

**Figure 2.3** Pre-post state histograms for resting in relation to feeding, at lag 0, in an ant. The solid and broken lines represent expected frequencies and 95% confidence intervals (from Douglas and Tweed, 1979).

ogy, like zoology in general, must inevitably deal with many aspects of an animal simultaneously, and hence its problems are inherently multivariate. As with any new means of travel, multivariate analysis takes a little getting used to, but affords otherwise unobtainable views for understanding behavioural dynamics. For instance, principal components analysis can extract dimensions of behaviour not readily apparent in the direct measurements made by researchers. A good example of this is the libido score of sexually active males of the smooth newt (*Triturus vulgaris*) (Halliday, 1976). This score correlates positively with the occurrence of a key display to females and with the total number of spermatophores released. Further, it declines with the depletion of sperm. These findings were similar in each male studied. The score thus proved to be a powerful index of sexual arousal in this species.

Correlational analysis can involve measures of association between the same or different acts, of varying proximity, within the same individual or between different individuals. Thus there are available intra- and inter- (cross-) correlations of different lags. Such tools are very powerful and sensitive but, for that very reason, must be handled with care. Many problems can arise; for instance, when observations are based on counts in successive time periods of fixed length, correlations are sensitive to this length. With brief periods, activities are often negatively correlated; at the other extreme, with long periods, positive correlations are found. As is true in all areas of science, inferring causality from correlations is hazardous. By its very nature, the temporal patterning of behaviour is also often non-stationary: the values of the determining factors vary over time. Examples discussed later in the Chapter will illustrate the contributions of these tools to motivational analysis.

Behavioural sequences can be viewed as a series of choices, and choices are of central interest to students of animal behaviour as well as those of human economic and political activity. Choices of, say, diet (Collier, 1982) are determined by both the cost and value of food items available in the environment and metabolic needs for calories and nutrients. An important set of derived measures for choice behaviour are preference metrics based on the relative counts of acts. Preferences exhibited by animals can provide important insights into underlying motivational mechanisms, with shifts in preferences indicating shifts in causal states. Various properties of preferences are of interest, such as their transitivity which refers to a preference for A over C when A is preferred to B and B to C. For instance, someone who prefers steak to hamburger, and hamburger to peanut butter will, under transitivity, prefer steak to peanut butter.

## Temporal analysis

The second route in ethological analysis has been to focus on how much time is spent in each activity, that is, the duration of intervals engaged in various responses of the repertoire. Such time budgets often serve as good initial descriptions of animal behaviour. Each interval is characterized by a time of onset and a time of offset. When it is of interest to examine how long an animal engages in a particular activity at a stretch, log-survivorship plots (which plot the logarithm of the probability of observing an interval longer than any value against that value) are often useful. If intervals are equally likely to terminate at any point in time, such plots are linear, and the number of intervals occurring in observation sessions of fixed length is described by a Poisson process. In this case, the intervals are statistically independent and exponentially distributed. Often such plots are not linear. Figure 2.4 provides an example of convex data indicating a clustering of acts over time.

If the exponential distribution does not provide a good description of the observed intervals, more complex distributions such as the gamma distribution, of which the exponential distribution is a special case, can be used. If responses occur independently at some fixed, mean rate, the exponential distribution describes the interval until the first response occurs, while the gamma distribution describes the interval until any number of responses occur. Additionally, the

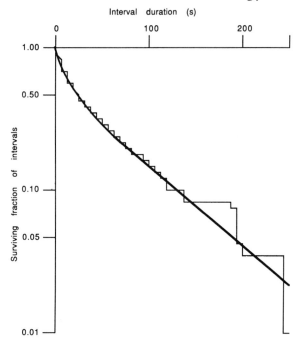

**Figure 2.4** Log-survivorship plot of intervals between mating in the fish *Barbus nigrofasciatus*. The surviving fraction of intervals longer than any value is shown as a function of that value (from Putters *et al.*, 1984).

dependence between successive intervals may be modelled. Animals often perform bouts of activities, producing clusters of brief intervals of activity separated by long, inter-bout intervals, and separating these intervals is problematical (Slater and Lester, 1982). If behaviour is hierarchically arranged, it would be expected that bouts of bouts should be observed, and indeed such super-bouts have been found in the pecking patterns of chicks (Machlis, 1977; Figure 2.5).

Temporal patterning has several further aspects. Operant psychologists have dealt with timing of responses in instrumental situations, generally from a non-motivational viewpoint. Many of these aspects are considered by Richelle and Lejeune (1980). The existence of rhythmic temporal patterns of behaviour leads to the study of biological clock mechanisms (Aschoff, 1981; Zucker, 1983). The periodicity of rhythms ranges from seconds (as in quick feeding and grooming motions) through hours (meal patterning, tidal changes) and days (reproductive cycles) to months (seasonal movements). These periodicities can be examined through the use of such techniques as spectral analysis, which reveals the relative importance of cycles of different

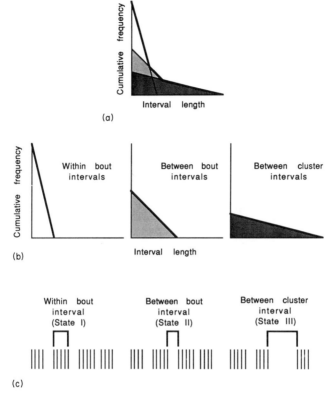

**Figure 2.5** Illustration of hierarchical, super-bout structuring of behaviour. The three behavioural states (c) each produce a log-survivorship plot (b) whose composite is shown in (a) (from Machlis, 1977).

periodicities from a spectrum of such components (Figure 2.6). These periodicities are approximate, as indicated by the prefix 'circa' in such terms as circadian and circannual, and are generally driven by endogenous oscillators. However, they do require setting by environmental cues, or Zeitgebers. In many situations behavioural rhythms and preferences interact in an intriguing manner. For instance, Bonsall *et al.* (1978) examined how preferences for males varied in female rhesus monkeys over their monthly cycle of receptivity. In the experimental situation, the females were required to make 250 instrumental responses in order to gain access to a male. Near the time of ovulation, females made the set of responses quickly for either preferred or non-preferred males, whereas at other times of the month they took longer for preferred males and even longer for non-preferred males. This is one of many cases in which

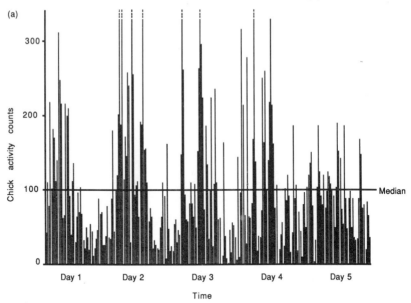

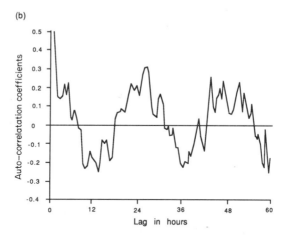

**Figure 2.6** Rhythmic activity in domestic chicks. (a) Counts of activity in successive 30 min intervals; and (b) auto-correlations and (c) spectral analysis of these counts (i) before and (ii) after removal of 24 h periodicity using multiple regression. Probabilities indicate the level necessary for the significance of spectral components (from Broom, 1979).

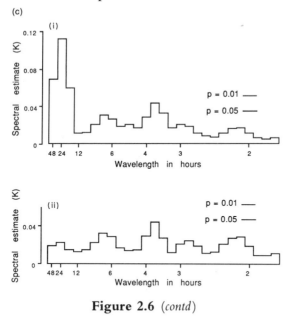

**Figure 2.6** (*contd*)

observed performance is the result of an internal rhythm interacting with external stimuli.

## Sequences, time, and the ethology of motivation

To obtain a comprehensive understanding of behavioural processes, it is necessary to consider both the sequential patterns and the temporal dimensions of activities. Motivational analysis involves studying changes in these patterns and dimensions which occur as the result of changes in external and internal states. A number of detailed ethological studies stand as examples of such analysis. For instance, in a series of papers Tim Halliday (e.g. 1975) elucidated how courtship in the smooth newt arises from the interaction of, on the one hand, three locomotory activities by the female (approach or withdraw from the male or remain stationary) and, on the other hand, the relatively rich display repertoire of the male. By focusing on the sequential and temporal aspects of the activities of the male, Halliday demonstrated both observationally and experimentally that this courtship behaviour is determined jointly by the internal state of the male, including his sexual arousal and his need for oxygen (since courtship takes place under water), and the reactions of the female to his displays. Further case studies of the interaction of internal and external cues will be considered at greater length in section 2.4.

The use of drugs is one way of experimentally manipulating motivational states. For instance, apomorphine, a dopamine agonist which leads to stereotyped behaviour, was administered by Lee Machlis (1980) at two dosage levels to three-day-old chicks pecking at red and white stimuli. Apomorphine treatment eliminated preferences between the two stimuli (Figure 2.7a) but temporal patterning still indicated discrimination between them by the chicks (Figure 2.7b). Thus this drug differentially affects preferences and temporal patterning.

The simultaneous examination of sequential and temporal patterning requires a top-down approach, which describes the operation of the entire system first and then identifies components. Such an approach, in contrast to a bottom-up examination of components which are later synthesized into a whole, has been widely regarded as unrealistic. However, Jan Leonard (1984) advocates top-down analysis as not only possible but necessary. She argues that particular attention must be paid to the superposition of motor patterns, i.e. the simultaneous occurrence of more than one pattern. Hence a complete record of behaviour is essential, and such a record is a time series which can be investigated by sequential, correlational, and spectral analyses. Using these tools, she concludes that, in jellyfish, the components of swimming bout, pause duration and multiple rhythms are subunits of the behavioural system producing swimming. It remains for top-down ethological analyses for species with more complex repertoires to be conducted. For instance, locomotion in vertebrates could be analysed from the bottom up as a sequence of muscular contractions recorded by a functional morphologist, and from the top down as approach to food or avoidance of a predator. Quantitative models useful in this context are considered in section 2.6.

The problems facing ethologists concerned with motivation can be posed as a series of questions dealing with ontogeny, dynamics, and frameworks. In terms of ontogeny, how does motivation develop in a growing animal? Learning is an ontogenetic process of particular importance in many species. How is learning affected by the motivational state of an animal? In terms of motivational dynamics, how does the impact of stimuli from the external environment combine with that of stimuli from the internal environment within a causal system? How does switching occur between causal systems, and what other sorts of interactions take place? In terms of frameworks, how can motivational processes be described quantitatively? And, finally, do cognitive concepts contribute to our understanding of motivation in such key areas of animal behaviour as social communication? These questions lead to the case studies

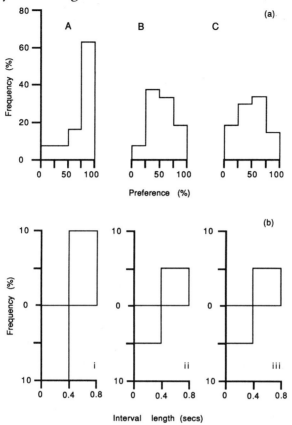

**Figure 2.7** (a) Distribution of preferences for the white stimulus in control (A) and apomorphine-treated (B 0.3 mg/kg, C 0.4 mg/kg) three-day-old chicks. Only the control chicks preferred this stimulus. (b) Similar distributions of relative numbers of intervals between two pecks at the same stimulus (R–R intervals minus W–W intervals) (from Machlis, 1980).

in this chapter: the ontogeny of feeding motivation in chicks; motivation as a constraint on learning; the summation of internal and external factors within a causal system; interactions between motivational systems (including the problems of conflict, displacement acts, and intention acts); quantitative models of motivational states; and finally cognitive approaches to animal behaviour.

## 2 ONTOGENY OF FEEDING IN CHICKS

Of Tinbergen's four focal points for ethology, ontogeny is the least studied, aside from specialized phenomena such as imprinting (see

Bateson and Klopfer, 1982). Ontogeny includes changes which are affected by specific experience (leading to learning and imprinting) as well as those changes resulting from the maturation of sensory, nervous, hormonal, and muscular systems, which are independent of such experience. Some ontogenetic processes continue over the lifetime of an animal. For instance, olfactory reactions in rodents depend in part on the maturity and gonadal condition of the animal. Other ontogenetic processes are shorter term, and include the priming effects of stimuli. Many ethologists, such as Tinbergen in his classical definition of instinct, have distinguished the priming and releasing effects of stimuli. Releasing effects combining with internal state are discussed in section 2.4. Motivational systems are said to be primed by appropriate stimuli when the arousal effects of those stimuli outlast their presence (Hogan and Roper, 1978). For instance, parental behaviour can be primed in diverse vertebrates by presenting eggs or young. Additionally, lactating guinea pigs, unlike virgins which are not primed by a maternal state, vocalize in response to the calls of infants. Priming is discussed further in relation to learning in section 2.3.

Behavioural development is especially clear in young animals in which growth and maturation are proceeding rapidly. An outstanding instance of such fast development is seen in domestic chicks (*Gallus gallus*) during their first ten days after hatching (Broom, 1968; 1980). During this period, chicks exhibit increasing frequencies of a number of activities involved in locomotion and feeding. Compared with birds raised in isolation, birds raised in groups show a similar course of behavioural development, but the changes are quicker. There are also rhythmic components in this development, of both daily and shorter periodicities. Additionally, the chicks change in the extent to which they react to alterations in their environment. For instance, both the magnitude and duration of reactions to changes in the level of illumination increase, especially during the first two to three days. The complexity of the rearing environment influences this reactivity. Chicks can be raised in simple environments, such as a pen with grey walls or with a view of stationary objects; or in more complex environments, incorporating moving objects or mirrors. After a change in illumination, six-day-old chicks in simple environments exhibit less locomotion and more calling compared with chicks in more complex environments. These results can be interpreted in terms of the roles of arousal and novelty in behavioural ontogeny. The chick develops an internal representation or model of its environment against which changes can be compared. The model is strengthened by longer experience so that older chicks detect novel changes more readily and show greater responses. A richer experience leads to greater ability to deal with novelty, so chicks which have had such experience are less disturbed by changes than are those reared in simpler environments. These investigations outline the ontogenetic background within which attachments are formed in social imprinting and motivational processes develop.

Jerry Hogan has focused on the ontogeny of the motivation for feeding in junglefowl chicks (*Gallus gallus spadiceus*) in a variety of experiments. In one set of experiments he presented young chicks with mealworms. Such chicks encountering mealworms for the first time are in conflict about how to deal with them. The mealworm can elicit feeding, withdrawal, or immobility reactions. If the chick eats the mealworm, it is very likely to eat subsequently encountered mealworms; if not, it is very unlikely to do so. Hogan investigated this conflict as an avenue to understanding the motivation of feeding in chicks. The activities, especially those involving the mealworm, were examined in an experimental situation in which age, deprivation, and physical and social environment could be manipulated (Hogan, 1965; 1966). Both the mealworm and an unfamiliar environment induced fear, and these effects were additive. This finding is consistent with that of Broom, concerning the effect of a changing environment on the subsequent behaviour of the chicks. As fear increased, the percentage time spent engaged in general activities such as moving, pecking, and preening decreased. Shrill calling increased and then decreased, while sleeping and sitting showed the reverse pattern (Figure 2.8). While highest levels of fear resulted in immobility, the closer the chick approached to the mealworm the more likely it was to withdraw. Contrary to much previous discussion of conflicts between approach and withdrawal, in these chicks fearful behaviour and withdrawal represent not aspects of a single unitary flight system but instead separate motivational systems which are actually mutually inhibitory, while approach is determined by stimulus intensity.

Support for this conclusion comes from several sources. Fear of the mealworm is recognized in a chick by its fixating the item. As would be expected with the existence of mutual inhibition of fearful behaviour and withdrawal, approach occurs more frequently during fixating than does withdrawal, the latter being caused by intense stimulation. Indeed, fearful and maintenance activities generally are mutually inhibitory, and this is the important conflict underlying the behaviour of the chicks. The 'food running' reactions of chicks which run about with mealworms in their bills before eating them also reflect the nature of the conflict (Hogan, 1966). The duration of such running is inversely correlated with the latency in picking up the mealworm (Figure 2.9). Also in agreement with Broom's

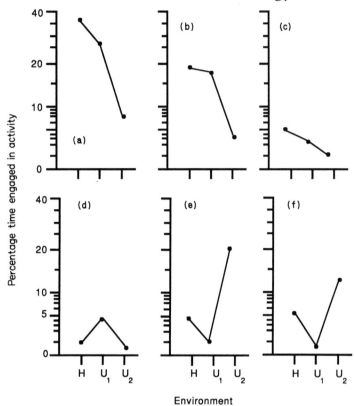

**Figure 2.8** Percentage time that young chicks engage in various activities as a function of fear which increases from the home (H) environment to a somewhat different environment (U$_1$) to a very different environment (U$_2$) Activities: (a) moving; (b) pecking; (c) preening; (d) shrill calling; (e) sleeping; (f) sitting (from Hogan, 1965).

results is the finding that, when first presented with a mealworm, week-old chicks run more than do younger ones. This can be interpreted as more fearful chicks hesitating longest and running least. Mealworms induce most fear on the first encounter, and thus chicks run about more on subsequent encounters. Contrariwise, decreased running is seen after an aversive experience with a mealworm which increases fear. Environmental unfamiliarity enhances hesitation and depresses running. Larger mealworms are more intense stimuli and so produce more withdrawal through running. Similarly, food deprivation facilitates running. All of these observations indicate the antagonism between fear and withdrawal. Food running wanes

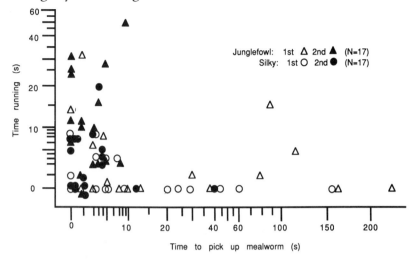

**Figure 2.9** Time spent running about with the first ($\triangle$, $\bigcirc$) or second ($\blacktriangle$, $\bullet$) mealworm in their bills decreases with the latency in picking it up in young chicks of two strains: junglefowl ($\triangle$, $\blacktriangle$) and silky ($\bigcirc$, $\bullet$) (from Hogan, 1966).

probably as a result of habituation to stimuli from the mealworm in the bill of the chick.

From an ontogenetic perspective, the motivational independence of fear and withdrawal is found only in young chicks. As the birds grow older, processes such as reinforcement probably link the control of withdrawal to that of fearful behaviour. Furthermore, the occurrence of emotional reactions is probably best described in young chicks by the James–Lange theory; that these reactions arise from the responses precipitated by external stimuli. For older birds a Cannon theory, that the reaction precedes the response, applies. (The James–Lange theory was put forth independently by William James and Carl Lange. Theories of emotion dealing with such issues as whether emotions cause actions or actions produce emotions are discussed by Young, 1961.) It is surprising that young chicks are very sensitive to changes in the external environment but not to food deprivation. Two-month-old chicks peck more in the absence of food when hungry than when not hungry and, when not hungry, peck more in the presence of food than in its absence. However, chicks one to two weeks old do not show these patterns. Perhaps feeding, like sexual and aggressive activities, has an independent causation early in the life of the chick and becomes linked into an integrated motivational system later.

This concept of the ontogeny of the feeding system was tested with a series of experiments on chicks over the first two months of life feeding on grain against either bare or sandy substrates (Hogan, 1971). Over the first two weeks of life, the various activities involved in feeding and locomotion are related to factors such as age, environment, and food deprivation in different ways. However, there are similarities among activities, as seen with pecking and ground scratching which both increase over time, and shrill calling and sleeping which decrease (Figure 2.10). Evidence for development comes from the interaction of age with other factors. For instance, pecking is initially most frequent in a sand-covered environment, but after the first two days is elicited chiefly by the presence of food and is less dependent on environment. Thus there is a change in the nature of the stimuli controlling pecking responses. The lack of effect of prefeeding chicks on their subsequent pecking and eating behaviour shows the initial absence of association between internal cues and feeding activity. The central activity of pecking is affected by several factors including the novelty of stimuli, drive as reflected in recent pecking, and, by the third day, the incentive value of stimuli and drive reduction as the result of experience. Initially, cues from the digestive tract play no role. Over the remainder of the first two months the association between pecking and ground scratching degenerates, contrary to the idea of a developing integration of activities. This degeneration may have occurred because under the experimental conditions, ground scratching did not function, as it does in nature, to uncover food, and hence was not reinforced by food.

The complexities of motivational development in the feeding behaviour of the chicks is clear from the diverse roles of different factors. Each activity has its own internal and external causes. Some activities, such as pecking and ground scratching, are linked even early in life, but since nutrition does not play a role at this stage, it seems inappropriate to invoke a hunger mechanism. Since correlations between pecking and weight change are low, different factors must control pecking and actual eating. Social stimulation from the mother hen as well as from social companions also influences the development of eating. Overall, contrary to the initial conception, chicks do not hatch with or develop a feeding system as some other vertebrates do. Perhaps, unlike their aggressive behaviour and sexual activity, feeding activity requires for its ontogeny functional experience of a type which could not be obtained in the experiments.

As part of a series of subsequent investigations into the development of food recognition in young chicks, Hogan (1977; 1984) has considered the importance of such functional experience in the ontogeny of the hunger system. In particular, he addressed the role

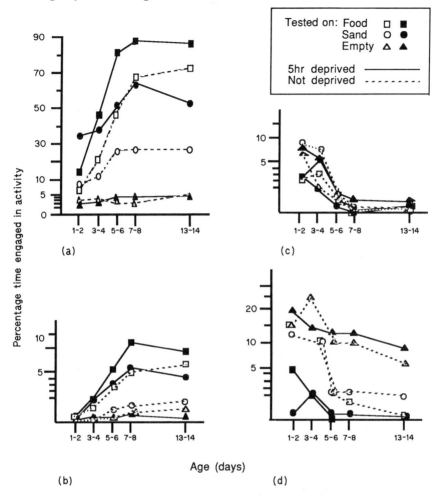

**Figure 2.10** Over the first two weeks of life, chicks (a) peck and (b) ground scratch more; and (c) shrill call and (d) sleep less as they grow older (from Hogan, 1971).

of associative and non-associative factors in the acquisition of discrimination between food and non-food items. Chicks were either force-fed with liquids or allowed to peck at food or sand prior to testing. Pecking rates at food and sand for ten minutes were then observed to see if a discrimination developed. The results, combined with those reported above, indicate that experience of pecking and ingestion, even if only with non-food items, is necessary for the discrimination between food and non-food by permitting the association of ingestion with its consequences. Pecking at sand glued to the floor, thus preventing ingestion, does not lead to a discrimination. The chief conclusion is that pecking and feeding are initially independent but after experience of pecking and ingesting pecking becomes linked with feeding, and deprivation effects are then seen on pecking rates. This is an instance of the general developmental principle that activities appear ontogenetically before they serve functionally. The lack of linkage between feeding movements and deprivation in early life is also seen in mammals, and highlights the need to separate causal and functional aspects of behaviour.

## 2.3 MOTIVATION AS A CONSTRAINT ON LEARNING

The topics of motivation and learning are, of course, closely linked. Learning alters the direction in which motivation takes behavioural performance, and motivation influences what is learned. In this section we examine the role of motivation in learning by reviewing the set of related processes which includes learning, and then proceed to the case study of motivational constraints in the ability of three-spined sticklebacks (*Gasterosteus aculeatus*) to learn certain tasks.

### Learning, priming and habituation

Learning is the chief process by which the effects of experience are recorded by animals. As such, learning has been and remains a major and diverse phenomenon of great interest. Much attention has been directed towards the mechanism(s) underlying learning. Reinforcement theorists maintain that learning occurs because of satisfying consequences of a response. These consequences are often viewed as involving drive reduction such as slaking a thirst. The term 'learning' is used by different people in different senses, but generally refers to a long-term change in the likelihood of a particular response following a particular stimulus over successive associations of the stimulus and response. Learning is usually explained as the outcome of classical and operant conditioning. Classical conditioning was first explored by the Russian physiologist I.P. Pavlov (1849–1936) who had become interested in behaviour while investigating digestive processes, including salivation, in dogs. With Pavlov's dogs, an initially ineffective stimulus, such as a tone, was paired with one, such as a puff of meat powder on the tongue, which is effective in eliciting the response of salivation. Over successive pairings, the tone came to elicit salivation when presented alone.

Against the long traditions of animal trainers, in circuses and on ranches alike, the first scientific studies of operant conditioning were carried out by E.L. Thorndike (1874–1949) with cats in puzzle boxes.

In operant conditioning, an animal performs a response, such as a bar press or running in a maze, which is instrumental in acquiring reinforcement. A great variety of reinforcement schedules have been developed by operant researchers. A reinforcement may occur after each response (continuous reinforcement, or CRF), after a fixed number, say 5, responses (fixed ratio 5, or FR5), after a variable number of responses averaging, say, 5 (VR5), after a fixed or variable interval of time (FI or VI), or after more complex combinations of these. It is traditional to contrast the involuntary elicitation of responses in a succession of fixed trials in classical conditioning with the voluntary emission of responses in an unrestrained manner in operant conditioning. However, the validity of this distinction remains unclear.

Learning experiments generally include two stages, acquisition, in which tone–puff or response–reinforcement pairings are maintained, and extinction, in which they are not and conditioning wanes. Learning researchers disagree over whether these conditioning paradigms reflect fundamentally different processes, and on whether all learning is based on these processes. Some researchers use the terms reinforcement or reward synonymously while others make a distinction based on classical versus operant conditioning, or aversive versus appetitive conditioning. There are also questions about the symmetry of the onset of aversive stimuli and the offset of appetitive ones, and the converse, as consequences of responses (Dickinson, 1980; Mackintosh, 1983)

As instigators to action, motivation and reward share similar functions and, by inference, probably similar mechanisms. Indeed some accounts invoke the incentive action of environmental stimuli in place of internal reinforcement, such as drive reduction, as the appropriate interpretation of learning. One such account has been put forward by Dalbir Bindra (1978). In contrast to the dominant theoretical viewpoint which emphasizes the association of responses and reinforcements, he has argued that links between perception and motivation are central to the problem of behaviour. In Bindra's view, motivation does not consist of internal organismic conditions but is the perceptual product of external stimulation by hedonic stimuli, including those termed releasers by ethologists. This account continues the historical controversy among learning theorists who regard learning as a formation of associations either between stimuli and responses, or between correlated stimuli alone.

Priming, discussed in the previous section as one major effect of stimuli, is also important in examining the motivational basis of learning. This is because both priming and rewarding properties of stimuli have effects on behaviour which outlast their presence,

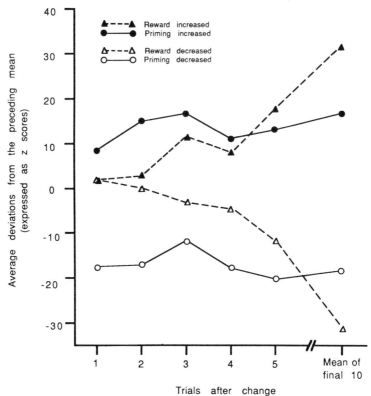

**Figure 2.11** Change in running speed over runway trials in brain-stimulated rats. Data are expressed as trial-by-trial deviations from prechange
mean (from Hogan and Roper, 1978).

but priming is regarded as a non-associative or motivational change
while reward is associative or structural. The performance in runway
trials of rats which are receiving electrical stimulation of the brain
illustrates this distinction. Such stimulation can be presented as
priming in the start box to the runway, and also as a reward in the
goal box. As seen in Figure 2.11, changing the priming stimulation
produces an immediate change in running speed which then remains
roughly constant, whereas changing the reward leads to a gradual
but much more prolonged alteration in the response. Incentive accounts of shifts in behaviour following changes in the level of reinforcement (e.g. Crespi, 1942) focus on priming effects of stimuli.
    A fundamental question in learning and motivation is the extent
to which different causal systems show similar dynamic properties
with respect to such processes as reinforcement. In a thorough com-

parison of reinforcers of different types, Hogan and Roper (1978) concluded that there are indeed some similarities in respect to the priming effects of stimuli and of magnitude and frequency of reinforcers during acquisition. However, there are also major differences across species, between the sexes, and in such features as the effects of deprivation and the extent to which compensation in response occurs when the effort needed to gain a reinforcement is changed. It seems that at present no physiological, behavioural, or evolutionary theory is sufficiently powerful to account for the diversity of available observations.

Learning must be distinguished from other processes which can change the probability of a response as a result of experience, including sensory adaptation, motor fatigue, and habituation. Habituation refers to the waning of responsiveness to repeated or constant stimulation caused by a central change in sensitivity. Phylogenetically, habituation is a very widespread phenomenon (Peeke and Petronovich 1984). Moreover, habituation, beyond being a process of interest in its own right, may methodologically confound studies in which stimuli are repeatedly presented to animals, as in the assessment of nest defence in birds (Knight and Temple, 1986). In the absence of knowledge about the underlying physiological mechanisms, it is not possible to know whether habituation (and its reverse, sensitization) differ fundamentally from learning.

## Constraints in sticklebacks

The constraints issue embodies the realization that learning is dependent not only on stimulus and response, but also on the reinforcer, species and individual, as well as the context, as discussed by Hinde and Stevenson-Hinde (1973). Arbitrary choices of these parameters will not necessarily lead to learning. Thus, analyses of the pertinent motivational mechanisms are needed. Piet Sevenster (1968; 1973) has provided a particularly clear example of constraints using sticklebacks, ethology's counterpart to the rat of psychology. The dramatic agonistic and reproductive behaviour of this species has been thoroughly investigated. Of particular note is the antagonism which exists between aggressive and sexual activity. While courting a female, a male performs a zigzag dance which involves alternating components of leading her toward the nest and approaching her. During this dance, the male exhibits a much reduced frequency of biting. Conversely, when aggressive responses dominate sexual behaviour, as occurs immediately after spawning, females are actually attacked. This antagonism serves as a constraint on the ease with which males can be conditioned.

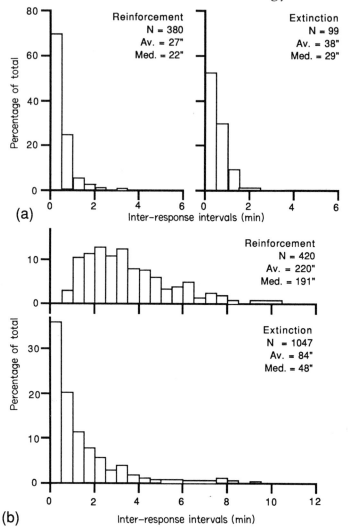

**Figure 2.12** Percentage distributions of inter-response intervals by male sticklebacks during acquisition and extinction, with presentation of a female as a reinforcer, with the response of (a) ring-swimming or (b) rod-biting (from Sevenster, 1973).

In his conditioning experiments, Sevenster employed, as responses, swimming through a ring or biting a rod and, as reinforcers, the presentation of a gravid female in a glass chamber for 10 or 20 seconds, or the similar presentation of another male. With the male reinforcer, the frequency of either response was enhanced, as expected according to standard conditioning theory. The interesting outcome occurred with the female reinforcer. With this reinforcer, males could be trained to swim through a ring, with most inter-response intervals (IRIs) (which include the presentation of the female) being less than a minute (Figure 2.12a). During extinction, the response rate initially decreases only slightly compared with that during acquisition. By contrast, when the instrumental response required was biting a rod, conditioning was slow and most IRIs were longer than two minutes (Figure 2.12b). The finding that the response rate during extinction greatly increased over that during acquisition indicates that the rod-biting is not limited to low rates of occurrence due to some sort of self-inhibition. Further, IRIs are similarly long after reinforcements and after presentations of the female independent of responses. Such presentations actually delay rod-biting. Additionally, after these presentations and after reinforcements, males exhibit zigzags before the rod, at a decreasing rate as time passes.

It seems that over the course of the rod–female pairings, the male comes to treat the rod as a female surrogate. Thus there is an intrinsic incompatibility between the required biting response and a reinforcement consisting of the appearance of a female, which has a motivational after-effect of inhibiting biting. Recent experiments (Sevenster and Roosmalen, 1985) have focused on glueing behaviour, in which the male spreads mucus secreted by the kidneys on to nest material. Once again, constraints between response and reinforcement are found: presentation of a female did not have a consistent effect on the rate of glueing. In general, responses and reinforcers cannot be arbitrarily paired: rather, attention must be paid to the underlying motivational systems (see Hinde and Stevenson-Hinde, 1973).

## 2.4 COMBINATION OF FACTORS

In this section we shall consider the combination of internal and external factors causing responses; the value of viewing these factors as concurrent variables; and the case of vacuum and other spontaneous behaviour. The results of research with cichlids will illustrate these processes of combination.

### Internal and external factors

Behaviour can be viewed as a joint function of internal drives and external stimuli, the latter acting as releasers for consummatory responses or incentives for appetitive responses. The contributions made by the various features of an external stimulus can sometimes

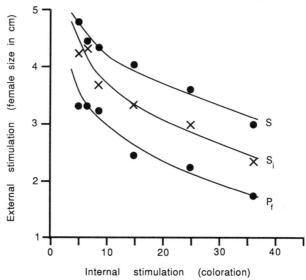

**Figure 2.13** Courtship in male guppies as the outcome of an external stimulus, female size, and internal arousal, as reflected in colouration, for posturing (Pf), sigmoid displays (S), and sigmoid intention displays (Si) (from Baerends *et al.*, 1955). See text for details.

be described by the **Law of Heterogeneous Summation** (Tinbergen, 1951) and can be investigated experimentally. In social situations, the stimulus instigating activity may be behaviour of conspecifics. In this context, two relevant phenomena are social facilitation and local enhancement. Social facilitation refers to the performance of a response, which is already present in the repertoire of the animal, due to such performance by others (consider contagious yawning). Local enhancement refers to an animal responding to some part of the environment in a manner influenced by the responses of others to this part. An example of this is avoidance of dangerous areas avoided by others. Researchers interested in learning in social situations must be careful to distinguish performance changes due to such learning from changes due to these motivational phenomena.

Sensory mechanisms transform external stimuli into signals in the nervous system which are integrated with messages from internal sources to determine outgoing responses. For instance, sexual behaviour in male rats can be viewed as the output of a system integrating gonadal signals with cues from females (Toates, 1980). This topic of summation has especially been studied with cichlid fish (see review of motivation in fish generally (Colgan, 1986)).

In their classic report on courtship in male guppies, Gerard Baerends and his colleagues (1955) presented what has become the foremost example illustrating the summation of internal and external factors to influence behavioural output. Males prefer large females over small ones, which is adaptive since brood size increases with female size. The arousal state of a courting male is reflected in coloured skin patches whose intensity is under neural control. (Animals which change colour, such as chameleons and particularly spectacular cephalopods (Wells 1978; Hanlon *et al.*, 1983) offer richly aesthetic material for students of motivation and deserve more study.) As shown in Figure 2.13, female size (the external stimulus) and internal arousal combine to determine the courtship acts of the male, with more intense acts requiring higher combinations of external and internal stimulation. The values on the abscissa were derived by averaging the relative frequencies with which the courtship acts were associated with each pattern of colouration.

## Pre-spawning behaviour in cichlids

Baerends has long pursued meticulous and unexcelled research in this vein in both fish and herring gulls (Baerends and Drent, 1982). A recent report (Baerends, 1984) on the organization of pre-spawning behaviour in a South American cichlid, the black acara (*Aequidens portaleguensis*), provides an excellent case study on the summation of factors.

Reproduction in cichlid fish can be divided into two patterns involving either oral incubation of eggs and young, or substrate spawning. In the latter case, preparation for spawning is relatively extended in time. During this period, the aggressive and sexual motivation of the partners adjusts as reflected in their communicatory and nest-related activities. The reproductive cycle involves establishment of a territory, pair formation, spawning, and parental care in which the young wrigglers are tended in a series of pits. In terms of a familiar computer analogy, Baerends' objective has been to study the hierarchically organized behavioural software or programs, founded on the anatomical and physiological hardware, by which pre-spawning activities are organized. From more than 100 elementary acts, he recognized 25 focal activities of the fish which served in the subsequent examination using cluster and factor analyses. The external environment was constant apart from the behaviour of the mate whose influence could be assessed by correlating the activities of the two individuals. Accompanying the overt behaviour of the fish are variable colour patterns which prove useful in identifying motivational states.

The analysis revealed the existence in both sexes of three clusters of acts which can be labelled agonistic, nest-digging, and sexual. Over the days prior to spawning, these clusters are successively dominant in the behaviour of the fish. The data indicate how the mate is a source of stimulation for each partner. Contingency tables of acts by one sex following acts by the other indicate that each type of act is associated with several different activities by the partner. Thus the identified clusters are not the result of shared responsiveness to the same releasing stimulus. Rather, activity in one fish tends to induce one of several responses belonging to the same cluster in the other fish. Since the same activities occur throughout the pre-spawning phase, the shift in dominance of the three clusters of acts is not due primarily to changes in external stimulation, but instead is due to motivational changes in the fish, with external cues playing a modulatory role. On the part of the female, an increase in reproductive tendencies leads to attempts to penetrate the territory of the male. Through her submissive but persistent behaviour, which Baerends terms **symbolic inferiorism**, she establishes herself in his territory. For the male, there is a gradual motivational shift from aggression to courtship, and consequent acceptance of her. Detailed examination of the data shows that males and females alter their motivational states in sex-specific patterns. The clusters of related acts reflect these states, and correlational analyses indicate that the states fluctuate in strength over a time course of a few minutes.

Conflict between motivational tendencies in cichlids, as in other species, is seen in the occurrence of various activities. The picking up of objects reflects partially suppressed feeding tendencies. Digging acts seem to have evolved from interruptive behaviour (discussed in the next section), which also includes cleaning acts such as finflicking and chafing. All of these behavioural patterns illustrate the widely found conflict between aggressive and sexual motivation in courting animals. Examination of the data for the black acara and comparison with findings for other species, including interactions of hormones and behaviour in birds, lead to the conclusion that the agonistic cluster contains opposing attack and escape components. By contrast, the cluster of sexual acts seems unified by underlying endocrinal machinery.

Courtship activities have evolved in the context of the conflict of agonistic and sexual tendencies, and can be interpreted as having a hierarchically ordered structure (Figure 2.14) as postulated by Lorenz and Tinbergen in their concept of instinct. At the highest level are the two overall categories of lifestyle, one involving reproduction and territoriality (RT) and the other involving a more pelagic (MP) existence. Two other high level systems, relatively independent of

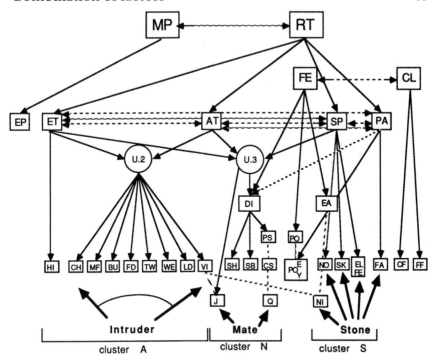

**Figure 2.14** Model of the organization of pre-spawning behaviour in the South American cichlid fish *Aequidens portaleguensis* (from Baerends, 1984). See text for details.

these two, are feeding (FE) and cleaning (CL). Third order elements include escape to the surface (EP) or to the substrate (ET), attack (AT), spawning (SP), and parental care (PA). The relative strengths of these elements are compared by units U.2 and U.3, while the focal acts (HI through FF) form the lowest levels of the hierarchy. Heavy arrows indicate external stimuli, continuous arrows stimulation, and dotted arrows inhibition. Feedback relations, not included in the diagram, also influence the dynamics of the hierarchy.

Baerends' research is an inspiration for students of ethology in its multifaceted contributions, i.e. its methodological scrutiny and thoroughness, its detailed and relevant interpretation, and its conceptual clarity on fundamental issues. Given Lorenz's reputed anti-quantitative bias, reflected in his pride at never having drawn a garden-variety X–Y plot, it is surely ironic that the most compelling support for his basic theory on the organization of natural behaviour

Ethology of motivation

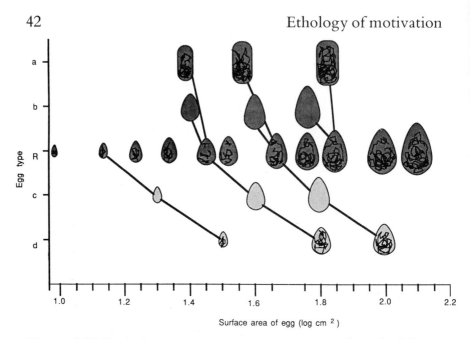

**Figure 2.15** Equivalences among egg surrogate stimuli with different features and surface areas. a: Brown, speckled, block-shaped; b: brown, unspeckled, egg-shaped; R: reference series of different sizes; c: green, unspeckled, egg-shaped; d: green, speckled, egg-shaped (from Baerends and Kruijt, 1973).

has been yielded up by a judicious employment of the most powerful contemporary multivariate statistics.

## Concurrent variables

It is worth noting that the study of the combination of drives and incentives has a long history. The topic has dominated much of learning theory within psychology. Within ethology, Baerends and Kruijt (1973) have developed what they have termed **titration** in order to assess the relative strengths of different stimulus features. They have particularly pursued this approach with respect to the roles of size, colour, and position cues of egg stimuli in eliciting retrieval responses from nesting herring gulls (*Larus argentatus*) (Figure 2.15). Such work is clearly affiliated to studies of preference as the outcome of evaluations along several dimensions. For example, the choice of a large hamburger now versus a small steak later involves considerations of the size and quality of food as well as the temporal discounting of value.

An important activity in this sort of study is the search for concurrent variables.

A variable is said to be concurrent if the ordering of the responses that arise from changing the value of the variable is the same whatever the value of the other variables may be (the values of the other variables are held constant during each determination of an ordering) (Houston and McFarland, 1976).

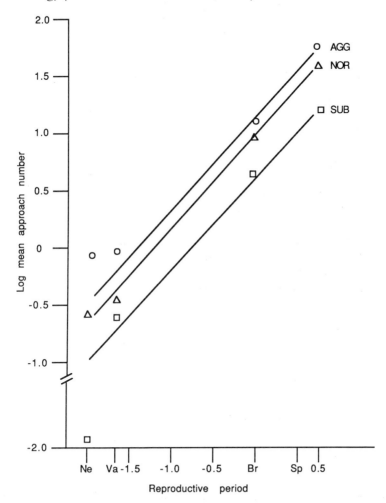

**Figure 2.16** Approaches by male pumpkinseed sunfish are an additive outcome of the external stimulus (the posture of the stimulus male surrogate: AGG: aggressive, NOR: normal, SUB: subordinate) and the internal state of the test fish (as reflected in its reproductive period: Ne: nesting, Sp: spawning, Br: brooding, Va: vacating) (from Colgan and Gross, 1977).

The variables in question may be internal or external, and appropriate scaling can often produce additivity. For instance, in nesting male pumpkinseed sunfish (*Lepomis gibbosus*), aggressive responses are the joint outcome of two concurrent variables (Colgan and Gross, 1977). The first variable measures internal state as reflected in the period within the reproductive cycle. The four periods of the cycle are nesting, spawning, brooding, and vacating. During these four periods the male establishes a nest, spawns with one or more females, cares for the developing brood, and finally vacates the nest. The second variable measures external stimulation due to presentation of stimuli resembling conspecifics in different postures: aggressive, normal, or subordinate. As Figure 2.16 shows, aggression is the additive result of the posture of the stimulus male surrogate and the period of the reproductive cycle of the test fish. For each period, the ranking of the three stimulus surrogates is the same, and similarly for each stimulus the ranking of the four periods is the same. The parallelism of the lines indicates the additivity of these internal and external factors.

Walter Heiligenberg (1976) has also demonstrated the additivity of factors in the aggressive behaviour of both cichlids and crickets. Figure 2.17 provides an example of the effect of presenting conspecific surrogates to males of an African cichlid, *Pelmatochromis kribensis*, for 30 seconds. Part (a) of the figure shows the number of attacks around the time of presentation of the surrogate averaged over 150 trials. Clearly the number increases sharply immediately after presentation and then decays exponentially with a half-time of about 1.5 minutes. Panel (b) plots the mean number of attacks in the first minute after presentation against the number in the final pre-stimulus minute when the surrogate was presented (open circles) or withheld (solid circles) in control trials. The activity of the fish is neatly seen to be the joint outcome of an external stimulus adding to the current internal state. Quantitative aspects of the response can be easily modelled by analog (continuous, as opposed to digital) control systems containing feedback loops and postulated elements termed state variables with appropriate decay constants.

## Vacuum activities

The role of external and internal factors is central to the issue of vacuum activities. Such activities were defined by Lorenz as responses occurring due to an accumulation of drive for that response in the absence of the usual adequate stimuli. Lorenz, as well as other European researchers such as the British ornithologist David Lack (1910–1973), cited various instances such as aggression by dogs and prey capture by birds, and attributed these actions to the accumu-

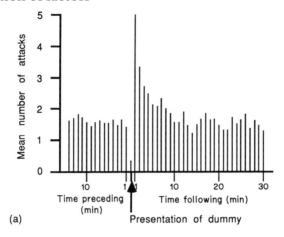

(a)

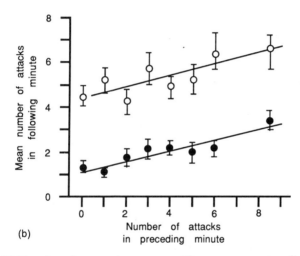

(b)

**Figure 2.17** Results of presenting conspecific surrogate stimuli to males of the African cichlid *Pelmatochromis kribensis*. (a) Attack rate before and after presentation averaged over 150 trials; (b) attack rate following presentation plotted against prior rate for trials including (open circles) or omitting (closed circles) the stimulus (from Heiligenberg, 1976).

lation of **action specific energy**. The (in)famous hydraulic, or flush toilet, model of instinctive motivation proposed by Lorenz was based on this energy concept. (Non-European animals are apparently more restrained.) It is not clear how to investigate experimentally the phenomena reported in these anecdotes. They represent extreme cases of the joint production of all responses by internal and external factors which must be uncovered in each case. All spontaneous behaviour poses this same problem for analysis. While some responses occur only under very specific conditions of external stimulation, vacuum responses, like exploratory activities, exhibit great spontaneity. Between these two extremes, most responses show the influence of both internal and external stimuli. Furthermore, the accumulation of action specific energy has little empirical basis. The topic of behaviour in the absence of obvious eliciting stimuli has often been treated in psychology in terms of frustration and thwarting, instances of conflict which are dealt with in the next section.

## 2.5 INTERACTIONS BETWEEN MOTIVATIONAL SYSTEMS

To deal with interactions between motivational systems is to come to the heart of causal analysis where the problems are both most fascinating and most baffling. In this section we shall consider the problem of interactions between motivational systems: the occurrence of conflict, the nature and causation of displacement activities, and proposed mechanisms underlying behavioural switching. The problem of finding a common currency for the apples and oranges of different motivational systems is a central one, and has long been obvious to investigators. In a cogent analysis of this issue, the Scottish philosopher Thomas Reid (1710–1796) queried

> I grant, that when the contrary motives are of the same kind, and differ only in quantity, it may be easy to say which is the strongest. Thus a bribe of a thousand pounds is a stronger motive than a bribe of a hundred pounds. But when the motives are of different kinds, as money and fame, duty and worldly interest, health and strength, riches and honour, by what rule shall we judge which is the strongest motive?

The need for animals to establish behavioural priorities is as clear to novelists as to philosophers and ethologists. A literary example occurs in Hugh Maclennan's *Barometer Rising* in which the physician says to the heroine about her recuperating fiancé 'He needs all the sleep he can get. After another few hours he'll wake up hungry. And then he'll eat like a horse. After that he'll want you. If you take my advice, you'll leave everything to him. He's got survival value'.

In one of his many apt metaphors, Lorenz (1966) has referred to the Great Parliament of Instincts:

a more or less complete system of interactions between many
independent variables, its true democratic nature has developed
through a probationary period in evolution, and it produces, if not
always complete harmony, at least tolerable and practicable
compromises between different interests.

The stream of behaviour constantly reflects choices among options
and thus close attention is warranted to the conflict among causal
systems for the **behavioural final common path**. This term was
introduced by McFarland and Sibly (1975) as the ethological equi-
valent of Sherrington's famous concept, referring to the final route of
activation among rival reflex arcs.


## Conflict

The expression of any one system is likely to be interfered with by
other systems. The ubiquity of conflict in motivation has long been
appreciated in both ethology and psychology, where accounts such
as that of approach versus withdrawal by N.E. Miller (1971) have
been influential. Conflict can arise when incompatible tendencies are
simultaneously aroused, or when behaviour is thwarted or frustrated due to an inadequate stimulus situation or the inability to
perform responses appropriate to some stimulus. Since many social
situations involve conflicting motivation, and emotions are often
involved with social signalling, students of motivation and social
communication share joint interests as discussed further in section
4.5.
  While natural selection might be expected to minimize conflict in
motivation, ethologists focus on this topic because it provides insight
into the motivational mechanisms underlying the establishment of
behavioural priorities. From a viewpoint of motivational analysis,
behavioural consequences of conflict are seen in a variety of actions,
many of which are compounds of separate acts. These actions may
be partially inhibited (intention movements), ambivalent, or alternating. Actions which show a constancy of form independent of
frequency were described by Desmond Morris (1957) as exhibiting
**typical intensity**. In the case of such actions with fixed forms, the
level of motivation can be indicated by frequency, duration, or intensity. Many threatening and courting actions exhibit typical intensity,
and the function of such constancy is discussed in the context of
the evolution of signals in section 4.5. Finally, autonomic responses
are also often seen in vertebrates in conflict situations. Such responses
include colour changes, sweating, vasodilation, and piloerection.
They reflect the role of the autonomic nervous system in affective

responses, as acknowledged in the theories of emotions developed by William James and Carl Lange, and by Walter Cannon (Young, 1961).

## Displacement activities

'Displacement activities', a term coined by the British avian etho-logist E.A. Armstrong (1900–1978) in 1947, refers to acts which appear to be out of context or functionally irrelevant (Zeigler, 1964). The very label 'displacement' implies an alternative outlet for beha-vioural energy blocked from its normal course. Since an energy concept is not well supported, there is force to the argument that the label should be abandoned. Such **interruptive activities**, as they are neutrally labelled by Baerends and Drent (1982), may in fact be instrumental or communicatory, or they may lack any apparent function. Examples include body care or feeding acts during aggres-sive or courtship encounters, and these probably represent a causally heterogeneous grouping. Such activities tend to be ones with simple forms and high probabilities of occurrence in the behavioural reper-toire. External stimuli influence which particular responses occur. For instance, if food is present, an animal in conflict may snatch a bite to eat. It is not surprising that grooming is often seen in such con-texts since stimuli from the integument of an animal are omnipresent. Additionally, displacement responses tend to be hurried and incom-plete compared with the occurrence of the same acts in their normal context. Note, finally, that the displacement responses of psycho-analysis, such as the reaction of an individual who is frustrated by his boss and so kicks the cat, are termed **redirected responses** in ethology.

Several explanations have been put forth for the causation of dis-placement responses. One of these asserts that frustration from thwarted drives in conflict may overflow into other systems. Defining the precise meaning of the key terms in this argument is problem-atical. The disinhibition hypothesis has been developed to account for the occurrence of displacement responses and conflict activity more generally. Pavlov, who interpreted his results on classical con-ditioning in terms of processes of excitation and inhibition, defined disinhibition as inhibition of inhibition. For instance, a noise in the laboratory during the extinction stage of an experiment could rein-state a conditioned response; the noise had inhibited the inhibition of extinction on the response.

Under the disinhibition hypothesis for displacement responses, as developed by Iersel and Bol (1958) in their studies of preening in terns, two major causal systems, such as aggression and brooding,

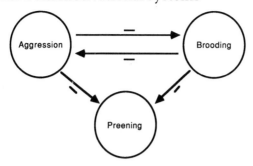

**Figure 2.18** Mutual inhibition of aggression and brooding, and inhibition of preening by each of these systems in terns (from Iersel and Bol, 1958).

are inhibitory on each other and on a third system, such as preening activities (Figure 2.18). These inhibitory relations could be demonstrated by appropriate manipulations of drive strengths through external stimuli such as conspecific males and females. An appropriate balance between the aggressive and brooding systems could lead to mutual inhibition of each other which removes the inhibition on the third system, and so displacement responses occur. According to this hypothesis, the motivation for these responses is autochthonous, that is, originates within that causal system. This assertion contrasts with the earlier idea that the motivation is allochthonous, overflowing from the frustrated system.

According to the disinhibition hypothesis, the intensities of the conflicting systems should not change the intensity of displacement responses, whereas under a frustrational overflow hypothesis these intensities could be related. Tim Roper (1984; 1985) examined these contradictory predictions with rats. These animals, some of which had previously had access to plain or flavoured water, were water-deprived to varying extents, and no water was available during testing. Compared with control animals, deprived rats devoted more time to behaviour directed towards the water spout, and all activities were speeded up. These effects increased with greater palatibility of the training drink and with increasing deprivation. Figure 2.19 shows the percentage time spent in various activities by the animals, and illustrates the increase in spout-directed behaviour by thwarted rats.

Roper regards the frustration model as unsatisfactorily vague, but his results also require amendment of the disinhibition hypothesis if it is to be maintained. Similarly, the development of the enhanced vigour observed in activities induced by mild peripheral stimulation, or under intermittent operant reinforcement or electrical stimula-

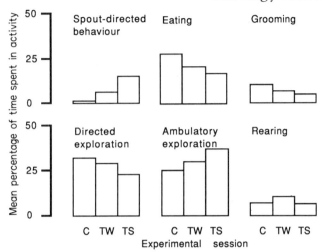

**Figure 2.19** Percentage of time spent in various activities for rats which had been trained with an empty spout (C), water (TW), and saccharine/glucose solution (TS) (from Roper, 1984).

tion of the brain, refutes the disinhibition hypothesis as applied to these situations (Roper, 1980). A third possible explanation of the increased vigour of displacement responses is in terms of post-inhibitory rebound, in which the occurrence of a response is enhanced after a period of inhibition. This phenomenon was first described by Sherrington as a spontaneous property of reflexes, and has also been observed in ethology. However, as Roper points out, such rebound is not seen in normal transitions from one activity to another. Hence the causation of both normal and displacement activities requires further study, with particular attention directed towards the possible existence of general facilitory processes which yield non-specific arousal across different causal systems.

In some situations, the occurrence of displacement responses may actually modulate the motivational state of the animal. This effect was seen in experiments conducted by Wilz (1970a; 1970b) on displacement responses in nesting male sticklebacks. In these males, the tendencies to lead the female to the nest and to chase her from it are in conflict. The ensuing zigzag dance by the male between the nest and the female is one of the most famous activities in the whole of ethology. In Wilz's work, the reactions of males to a female surrogate were recorded (Figure 2.20), with the aggressive level of the male subjects being manipulated by presenting them with a stimulus male in a bottle. Activities which were relatively common prior to stimulus presentation, such as pushing and bringing ma-

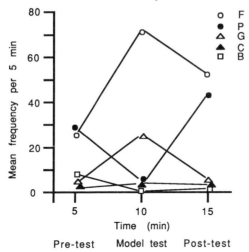

**Figure 2.20** Frequencies of nest-related activities in male sticklebacks presented with females: F: fanning, B: bringing material, G: glueing, P: pushing, and C: creeping through (from Wilz, 1970b).

terial, dropped in frequency during courtship, while glueing and fanning increased markedly. These results therefore refute the possibility that low threshold, or prepotent, activities are the ones which occur most often during conflict.

Particular attention was paid in these studies to dorsal pricking, in which the male nudges the female surrogate towards the surface of the water with his dorsum. Wilz showed that dorsal pricking tends to occur in response to an approach by a female when the male is aggressive or when sexual stimulation is lacking. Dorsal pricking causes the female to wait, while the male goes off to the nest, performs displacement activities (creeping-through and fanning), and returns to court. Females failing to wait are generally attacked, and attacks also result if the displacement activities are blocked. Thus dorsal pricking provides the male with an opportunity to perform displacement responses and thus to switch from primarily aggressive motivation to primarily sexual motivation. Wilz argues that such self-regulation of motivation is the function of these responses and that nest- and female-related activities are causally linked rather than antagonistic as previously thought. Subsequent research suggests that populations of males differ in the amounts of zigzagging and dorsal pricking which they exhibit when confronted with females, and that these differences reflect varying ecological conditions.

These various results illuminate the causal heterogeneity of displacement responses as a group. Because of their conspicuousness and diversity, displacement activities offer important material for the analysis of causal mechanisms of behavioural patterning.

## Mechanisms of behavioural switching

What proximate mechanisms could generate switches from one activity to another? Richard Dawkins (1976) has suggested that random switches occur between hierarchically organized systems to generate behaviour, and has outlined a clustering technique which focuses on the mutual replaceability of acts in sequences. The available data, from damselfish and great tits, do not support the proposal of random switches (Colgan and Slater, 1980). Furthermore, clustering by mutual replaceability is a useful concept for establishing the contextual similarities of acts but does not identify similarities based on common causal factors.

David McFarland (e.g. 1971) has attempted to embed the observations on displacement responses and the concepts of attention and disinhibition within a general framework for considering mechanisms responsible for establishing behavioural priorities and switching. In particular, he has elaborated three mechanisms, competition, chaining, and time-sharing. In competition, increases in the causal factors of a second activity oust an ongoing activity; in chaining, decreases in the causal factors of the current activity disinhibit a second activity, with successive decreases resulting in a chain of behavioural switches; and in time-sharing, or what could also be labelled motivational dominance, an ongoing activity remains dominant but terminates itself from time to time and so disinhibits the second, subordinate activity.

The metaphor of time-sharing is a product of computer jargon, and McFarland has executed a number of related studies in examining it. In one such study, the mechanisms establishing the priority of feeding and sexual behaviour in male rats presented with food and a sexually receptive female have been investigated by Brown and McFarland (1979). The latency, duration, and frequency of various activities involving eating food, mounting the female, intromitting, and ejaculating were recorded. In order to vary the level of the causal state associated with feeding, the male rats had been deprived of food for 0, 24, or 48 hours. The results indicate that such food deprivation does not affect temporal aspects of sexual or feeding behaviour. The authors therefore conclude that time sharing (rather than competition or chaining) was operative, with sexual behaviour dominant over feeding behaviour. This conclusion may be correct, but the report would be more convincing if information about individual performance, especially the number of ejaculations per session, was included. Additionally, since time-sharing is a hypothesis about the allocation of time to various motivational systems (in this case feeding and sexual behaviour), it would be helpful to see these results. On this point, some of the data presented, involving the time out between mounts and the time between successive mounts, appear to be influenced by food deprivation, and therefore suggest that competition may be occurring.

As a further test for time-sharing, McFarland has suggested interruption experiments, in which behaviour is interrupted by some method which is assumed to be motivationally neutral, that is, does not affect the levels of any causal factors. (Such interruptions have broad potential use in the study of motivation as indicators of the persistence and stability of activities.) When interrupted, dominant activities should be resumed but for subordinate ones this is not necessarily true; after a short interruption the subordinate activity may be resumed but after a long one the dominant activity is expected to have again exerted control. The length of the interruption can thus be titrated to assess the duration of the interval for which the dominant activity permits the subordinate one access to the behavioural final common path. Overall, interruptions should decrease the amount of time spent on the subordinate activity but leave that devoted to the dominant activity unaltered. Houston (1982) has challenged the underlying logic of time-sharing as a mechanism, and McFarland (1983) has replied, but matters remain unclear. Further on the structure of interruption experiments, it is worth noting the formal resemblance between such experiments and operant procedures involving temporal delays: in both situations an animal is unable to respond for some period of time. This resemblance suggests that it would be fruitful to examine behavioural priorities for various activities from both time-sharing and operant viewpoints by examining the extent to which animals persist in the same activity as opposed to switching to another.

Cohen and McFarland (1979) presented results from interruption experiments with nesting male three-spined sticklebacks as empirical support for time-sharing. Such males, who divide their time between caring for the nest and courting females, were kept in aquaria with three linearly adjoining compartments (Figure 2.21). They were allowed to nest in one end-compartment, and were presented with females in bottles in the other. Doors between compartments could be opened or closed to enable access or to interrupt ongoing behaviour. Dropping into the nests snails, which the fish removed, also interrupted behaviour. Principal components analysis of the activities of the male revealed three groups, associated with nesting, sex,

Ethology of motivation

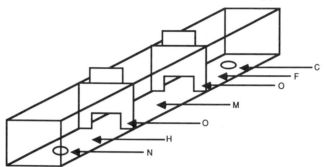

**Figure 2.21** Partitioned aquarium used to conduct interruption experiments as test of time sharing in fish. C: food cup; F: feeding compartment; H: location for intruder presentation; M: middle compartment; N: nest; O: opaque partition with sliding door (from Noakes, 1986).

and aggression, which compete for dominance. Seasonal variation in motivational levels was also noted by comparing fish artificially brought into breeding condition and fish breeding naturally. Consistent with the concept of time-sharing was the finding that the percentage of returns to dominant rather than subordinate behaviour increased with the length of an interruption.

It is not obvious that these results of Cohen and McFarland provide unambiguous support for the concept of time-sharing. Data were pooled across subjects and experimental sessions in which different activities may have been dominant. Over three-fifths of these data were discarded from the analysis. No control data were collected to establish baselines for the durations of visits for each subject. The relative dominance of the various activities was not established and alternative interpretations involving competition are possible. Since the doors between the compartments of the experimental aquarium were aligned, a male could see directly from the nest to the female. Nest- and female-related activities were therefore not independent. More generally, a function of displacement fanning may be to advertise the nest to the female. Finally, a recent replication, with improvements, of Cohen and McFarland's experiment by Stephen Crawford in our laboratory offers no support for the time-sharing hypothesis.

In a similar vein, David Noakes (1986) has examined how nesting male sticklebacks divide their time between feeding and nest activities. Instead of the females in the previous experiment, food was available for a portion of the day. Behavioural dominance could be influenced by manipulations such as deprivation of food or presentation of conspecifics. Such manipulation is essential in order to establish time-sharing. Interruptions were either 5, 15, or 30 seconds in length. Using a criterion of five or six returns to a compartment out of six interruptions, dominance could be established in 29 out of 76 observations, 24 for the nest and 5 for food. When subdominant activity was interrupted, short interruptions were less likely to make the male resume the dominant activity than were medium or long interruptions. By contrast, when dominant activity was interrupted, it was almost always resumed, regardless of the length of the interruption. Both of these results are as predicted by the concept of time-sharing. As expected, nest-related behaviour was greater during nest-dominant observations. The problem remains of how to account for the majority of observations in which there was no obvious dominance.

A final point on the suggested mechanism of time-sharing bears on its functional properties. McFarland has suggested that time-sharing may serve to ensure that an animal interrupts long bouts of feeding or drinking in order to scan the environment for potential predators or mates. Nonetheless, it is difficult to see why such a mechanism is superior to competition among causal states generated by the combination of external stimuli and internal drives. Such competition could include high-rate feedback mechanisms, with positive feedback preventing disadvantageous swift changes between activities, and negative feedback ensuring such changes when these are advantageous. The features of these feedback mechanisms would be selected in evolution to minimize the costs of meeting the needs of the various activities, and hence maximize fitness. This difficulty in the functional aspects of time-sharing will again be noted in section 4.1.

Much of the experimental work by McFarland's group has centred on feeding and drinking in doves (*Streptopelia risoria*). As a motivational model to deal with these activities, Lester (1984b) has suggested that the relative probabilities of the two activities should be proportional to the respective motivational levels taken as the product of deficit and expectation of reward availability. The expectation terms incorporate what the bird has learned from previous experience and thus reflect its knowledge about the environment. These terms are updated as feeding and drinking proceed and prime the motivational levels for these activities. Simulations based on these assumptions yield output like that seen in doves. This model is further discussed in section 4.1.

Finally, it should be emphasized that motivational systems are not fully independent of each other. This is clear in the nest- and female-related activities of three-spined sticklebacks studied by Wilz, and in breathing and courtship in males of the smooth newt (Halliday, 1977). Further, in this latter case transitions between activities may not be either exclusively competitive or disinhibitive. In this species, courtship is influenced by three factors: oxygen needs, sexual arousal of the male, and reactions of the female. Under an atmosphere rich in oxygen, there is competition between courtship and rising to the surface for a gulp of air, as indicated by variable intervals between breaths. However, under an atmosphere rich in nitrogen, the intervals are less variable, suggesting disinhibition of breathing through a rapid decrease in the causal factors associated with courtship. This conclusion, of course, requires support from experimental evidence in which such factors are manipulated. The broad implication of such a conclusion is that the search for the mechanisms underlying behavioural switches will require investigation of the activities of each species of interest. Since behaviour in general involves switching among a number of activities, this investigation will not be simple.

## 6 QUANTITATIVE MODELS FOR MOTIVATIONAL PROCESSES

Notwithstanding Skinner's approving recitation of the taunt that with an equation with three constants one can draw an elephant and with a fourth make him lift his trunk, various areas of applied mathematics have proven useful for generating quantitative models of motivational processes. Cybernetics has been employed here and elsewhere in animal behaviour (e.g. sensory and motor control, Barnes and Gladdon, 1985; orientation, Schone, 1984; feeding and drinking, Toates, 1980). With his work on courtship in glandulo-caudine fish, Keith Nelson (1965) was among the first to outline a quantitative model for temporal aspects of motivational processes. Using a cross-correlational analysis of activities, Heiligenberg (1976) showed that temporal patterning in the African cichlid *Haplochromis burtoni* under steady external conditions resulted from four basic processes. He thus provided for this instance an answer to the fundamental motivational question of how many causal systems are operative in a given situation. Beyond the work reviewed in the above section, McFarland and co-workers have elaborated ideas on motivational quantification (McFarland, 1971, 1974; Sibly, 1980; McFarland and Houston, 1981). More generally, modelling in this area is fundamentally linked to methodological and conceptual issues which are treated under psychophysical scaling. Psychophysics deals with the discovery of relations between dimensions outside an animal and their internal representations. In both psychophysics and motivational modelling researchers must choose both appropriate methods with which to study animals, and concepts with which to understand them. In this section, we shall examine the application of:

1. catastrophe theory to aggressive motivation in sunfish;
2. cybernetics to sexual motivation in rats;
3. statistical modelling of interactions between mothers and infants of rhesus monkeys.

## Catastrophic sunfish

Catastrophe theory presents excellent candidates as models for motivational analysis. This theory deals with the dynamics of systems that undergo apparently discontinuous changes (catastrophes). These catastrophes are viewed as occurring in a region termed the behavioural space, functions within which are controlled by continuously varying parameters in a region termed the control space. Many motivational situations, in which behaviour changes abruptly as the result of gradually changing underlying states (such as the arousal of cats discussed in Chapter 3), have this form. Catastrophe theory, a topic within differential topology, has been developed by the French mathematician Rene Thom who has shown in a long and difficult proof that for systems with not more than four controlling variables, there are only seven basic types of catastrophes (such as cusp and swallowtail, used in behavioural applications). The possible applications of this theory appear very diverse and worth investigating.

Catastrophe theory has been successfully applied to aggression in nesting male sunfish as they defend their nests against intruders (Colgan *et al.*, 1981). A cusp catastrophe with two control variables, reproductive period and distance to the centre of the nest, provided a good description of the results. Figure 2.22(a) shows how aggression and the two control variables are related by the surface of such a cusp catastrophe located in a three-dimensional space spanned by axes representing distance from the nest centre, reproductive period, and level of aggression. When the distance from the nest centre is small, the motivational state of the defending fish is represented by points on the right, higher portion of the surface, and aggression occurs. For longer distances, the state is represented by points on the left, lower portion of the surface, and the fish is not aggressive. The transition from aggression to non–aggression does not occur at the same distance from the nest as the reverse transition. This feature is termed hysteresis, and the magnitude of the difference in the distances for the two transitions is dependent on the reproductive period (Figure 2.22b). The data collected from large samples over three field seasons are well accounted for with such a cusp

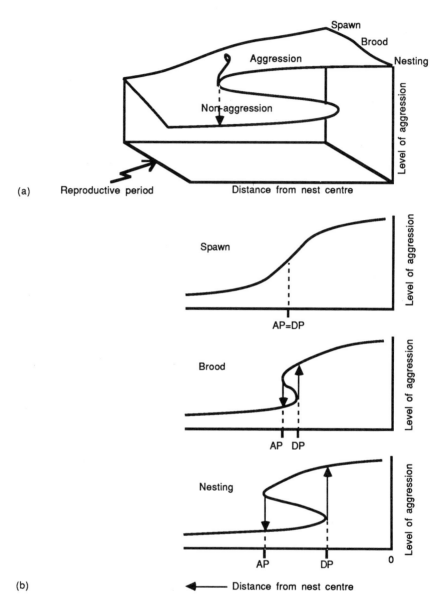

**Figure 2.22** (a) Cusp catastrophe relating behaviour (aggression and non-aggression) to the control variables (reproductive period and distance from nest centre). (b) The defence and attack perimeters (the distances from the nest centre at which aggression towards intruders begins and ends) vary with reproductive period (from Colgan *et al.*, 1981).

catastrophe, as well as providing information on the effects of size, posture, and speed of presentation of the stimulus surrogate fish. Probity forbids my dwelling further on this milestone of quantitative motivational research. Suffice it to quote Thom's summary of the importance of the field:

> Theoretical biology should be done in mathematical departments. We have to let biologists busy themselves with their very concrete – but almost meaningless – experiments.

## Cybernetic rats

A good example of the use of cybernetics in motivational analysis is the computer simulation of sexual behaviour in male rats carried out by Toates and O'Rourke (1978). (For readers concerned about sexual equality, the behaviour of females is discussed in Toates, 1980.) The simulation is in the tradition of physiological models of feeding and drinking. However, whereas these ingestive models rely on gross physiological measures such as gut and blood volume, a simulation of sexual behaviour requires motivational measures such as arousal and inhibition to model the pattern of intromissions which lead to ejaculation. The key elements of the simulation are shown in Figure 2.23. Potential arousal from an external source ($A_e$), the female, combines with potential internal arousal ($A_i$) to produce total potential arousal ($A_p$). Gain $K_1$ represents the state of arous-ability of the nervous system, with $K_1$ low in sexually exhausted males and high in eager ones. This gain $K_1$ operates on potential arousal $A_p$ to produce actual arousal ($A_a$). If this actual arousal exceeds the threshold for intromission ($T_1$), then intromission occurs. A second gain, $K_2$, relates intromission to the resulting neural excitation (N) which increases with successive intromissions. A third gain constant, $K_3$, incorporates the effect of neural excitation on internal arousal. When the actual arousal reaches the ejaculatory threshold ($T_2$), ejaculation occurs. Ejaculation decreases the arous-ability of the system (i.e. lowers $K_1$), initiates an absolute refractory period during which no sexual activity occurs, and sends an impulse to the inhibition integrator. This inhibition (I) decays over time, with a time constant represented by $K_4$, and copulation resumes when the difference between central arousal and inhibition exceeds the threshold for intromission.

A variety of experimental data, such as that from the investigation of the effects of the interruption of sexual behaviour after varying numbers of intromissions and ejaculations, support each aspect of this simulation. The simulation correctly models the patterning of sexual behaviour in which, for instance, fewer intromissions pre-

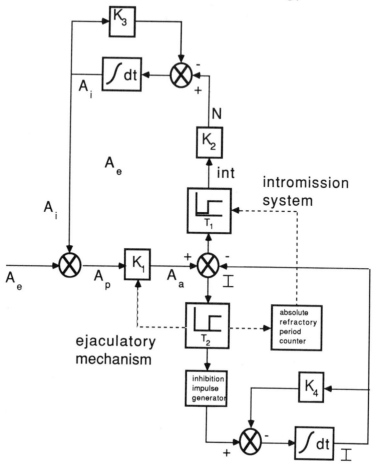

**Figure 2.23** Simulation of sexual behaviour in male rats (from Toates and O'Rourke, 1978). See text for details.

cede a second ejaculation as compared with a first. It also deals with phenomena such as the time course for recovery from sexual exhaustion and the stimulating effect of a novel mate. The model thus provides a very heuristic account of sexual behaviour in male rats; it also distinguishes explicitly among rival accounts of sexual motivation, and can be used to suggest further experiments.

## Markovian monkeys

Detailed models of behavioural processes have been particularly well developed by Dutch workers including Haccou *et al.* (1983); Metz *et al.* (1983); and Putters *et al.* (1984). One of these will serve as a case study. Efforts by Patsy Haccou and colleagues have resulted in both a sequential and a temporal analysis of interactions between mothers and infants in rhesus monkeys (*Macaca mulatta*). Observations made on pairs of animals alone or in the company of another pair indicated that the activity of the infants could be classified under three states: on the nipple; on the mother but not on the nipple; and off the mother. The research strategy was to break the data into successive periods, within each of which the probabilities of transition from one state to another are constant and depend only on the present state. Such periods are described by stationary Markov probability models. (Markov (1856–1922) was a Russian probabilist who developed the theory for these models.) The transition probabilities can be resolved into two components, reflecting acts by the mother and acts by the infant. The termination rate for any state is the sum of the transition probabilities to any other state. The assumption of a constant probability of a state ending can be examined with log survivorship plots.

The periods were identified by visual scanning of the data as well as by formal tests. Visual scanning consisted of bar plots of the occurrences of acts over time; plots of cumulative bout lengths of each state against the sequence number of the bout; and plots of log bout length against time (Figure 2.24). The latter plot, under the above assumptions, should be constant over time with a constant variance. These visual tests can be helpful, but can also be misleading: computer simulations show that such tests may indicate multiple periods where there is only one. Hence, formal tests, consisting of maximum likelihood estimation of the data reflecting a single period, two periods, and so forth, are essential. The power of these tests (i.e. the probability that they will reject a false null hypothesis) can also be estimated. For the special case of contrasting the hypothesis of a single period to that of two periods, a powerful test is available (Haccou and Meelis, 1986). The outcome of this analysis reveals how mother–infant interactions proceed as a series of periods, each characterized by particular levels of mother and infant involvement. The synchrony of the activities of infants was also made clear.

The simplicity and elegance of the description of these behavioural states is striking. The next step is to provide a similar analysis of the full ethogram. This is difficult for several reasons. Temporally, MAPs range heterogeneously from near point events to acts of long duration. In many cases (even in simplified systems such as fish feeding in a standardized environment, as studied in my own labora-

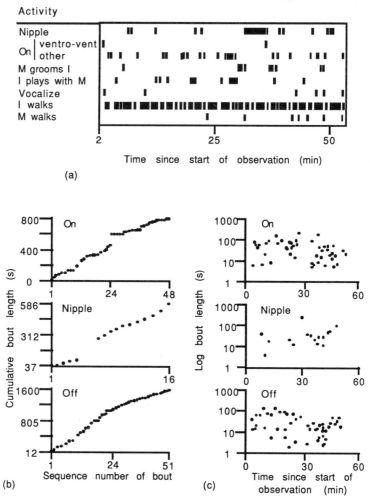

**Figure 2.24** Visual presentations of mother–infant interactions in rhesus monkeys. (a) Occurrence of various acts over time indicated by bars; (b) plots of cumulative bout lengths of each state against the sequence number of the bout; and (c) plots of log bout length against time (from Haccou *et al.*, 1983).

tory), the distribution of intervals is not exponential or even gamma. Instead, there are gradual changes in parameter values as, say, hunger decreases during a meal (unlike the stationary periods in the monkey study), and there are strong sequential dependencies between acts. The challenge remains to develop adequate descriptions of such behavioural systems.

## 2.7 COGNITIVE APPROACHES

The application of cognitive concepts to animal behaviour has gained momentum in the past decade (see Hulse *et al.*, 1978; Roitblat, 1987). Much of the momentum has arisen as opposition to the previously dominant framework of behaviourism. The latter seeks operational analyses of behaviour, that is analyses solely in terms of observable measures. Intellectually, behaviourism belongs to the tradition of logical positivism and operationalism. Logical positivism, developed by the famous Vienna Circle in the 1930s, sought to establish a scientific vocabulary consisting only of empirical and mathematical terms (see Achinstein and Barker, 1969). Akin to logical positivism, operationalism was espoused by the physicist P.W. Bridgman (1882–1961) who maintained that meanings are synonymous with operations (Bridgman, 1950). In psychology, positivism and operationalism led to the revolution of radical behaviourism, effected by John Watson (1878–1958), which limited discussion to observable behavioural units (e.g. Watson, 1930). Among contemporary psychologists, B.F. Skinner and his disciples have hewed most closely to Watson. In its treatment of data, ethology, like psychology, has been dominated by behaviourism. In the search for methodological purity, one of the most influential distinctions in psychology has been that between intervening variables, which are drawn from relations among empirical data, and hypothetical constructs, which are derived from theoretical considerations (MacCorquodale and Meehl, 1948). In a recent examination of the fundamental frameworks of Skinnerian and cognitive psychologists, Ben Williams (1986) has persuasively argued that this distinction cannot really be maintained and that both camps employ theoretical concepts. He suggests that behaviourists tend to be much more parsimonious, and that parsimony is facilitated by studying simpler problems. Certainly cognitive psychologists permit themselves richer intellectual landscapes than do spartan Skinnerians.

In contrast to behaviourism, cognitivism focuses on the internal representation of knowledge, and insists that theoretical concepts such as attention, expectancies, images, intentions, goals, plans, and templates are essential for the understanding of behavioural patterning (Figure 2.25). Just as Lorenz has coloured the ethological literature with his descriptions of, say, jealous geese (and produced controversy over the role of analogies in biology: see J. Cohen, 1975), so psychologists have often used emotional terms to describe putative cognitive features of their subjects. Even after the revolution of radical behaviourism, Crespi (1942) spoke of elation, eagerness and depression in describing the performance of his rats after changes

"Stimulus, response! Stimulus, response!
Don't you ever *think?*"

**Figure 2.25** (from Gary Larson's *The Far Side*, Universal Publishers, Chicago 1986).

in the level of operant food reinforcement. Similarly, Mowrer (1960) described the results of the onset and offset of attractive and aversive stimuli in terms of fear, relief, hope, and disappointment. Contemporary cognitivists no longer lurk in closets shamed by the glare of an ultramontane behaviourism, or wince at the criticism that a cognitive approach leaves an animal buried in thought and immobile in the midst of its world.

The lineage of hedonic psychology with its emphases on the emotions, as most fully presented by Young (1961) and Cabanac (1979), is also a major contributor to this blossoming of cognitivism. Hedonic processes have been assumed to underlie the central processes of reinforcement and incentive stimulation by various theorists (e.g. Solomon, 1982). Cognitive psychology has been further boosted by the growth of problem-solving algorithms in the field of artificial intelligence (e.g. Boden, 1981) and of cybernetic models which include such concepts as *Sollwert*, or set-point at which behaviour maintains an equilibrium. Cybernetics provided a chief impetus for that classic in cognitive psychology, *Plans and the Structure of Behavior* by George Miller, Eugene Galanter, and Karl Pribram (1960), in which a plan was defined as 'any hierarchical process in the organism that can control the order in which a sequence of operations is to be performed'.

Developments within an adaptationist framework (discussed in Chapter 4), such as the concepts of evolutionarily stable strategy and rules of thumb by which animals approximately optimize behaviour, have also been conducive to a cognitive approach. Even operant conditioning of stickleback responses (Sevenster and Roosmalen, 1985) and learning of foraging responses in Atlantic salmon (*Salmo salar*) (Marcotte and Browman, 1986) are dignified as examples of cognition when in fact simple accounts in terms of familiar, basic mechanisms are sufficient for explanation. Continuing in this same vein, Wyers (1985) has argued for a cognitive framework for the analysis of stickleback behaviour. The six aspects which he discusses (decision hierarchy; self initiation and goal orientation; means activities and least effort; temporal frame; spatial frame; and novelty and integration) are all of central motivational interest, but no cogent need for cognitivism is advanced.

Within ethology, arguments for a cognitive approach, most notably from Donald Griffin (1984), have led to a loosening of the strict methodological emphases of operationalism. Griffin and kindred spirits are impressed that the richness of communication and activity in a great diversity of species must reflect rich mental experiences. It is certainly true that such communication is more complex, both in the variety of modalities used and the elaboration within each of these, than was realized two decades ago. It is also parsimonious to infer that there is mental continuity across species because there is neural continuity. In particular, the investigation in humans of event-related potentials, i.e. neural signals associated with various cognitive tasks, should certainly be extended to other species. However, Griffin's conclusion that through animal communication we thereby possess windows into animal minds is opaque. His vague and unsatisfactory definitions of mental events such as awareness and mind indicate a critical lack of the necessary operational basis for a new scientific endeavour which is to venture where behaviourism has failed to tread. (Behaviourists who are foolish enough to attempt to define terms such as consciousness and mind (e.g. Rachlin, 1970) fare no better. One should not try to utter the ineffable.)

In the actively researched area of vocal learning in birds, Peter Marler's (e.g. Marler and Sherman, 1983) concept of an underlying neural template which is shaped by experience and guides the production of song is a prominent cognitive concept: 'the learned memory is used template-fashion as a reference for song development'. Concurrently, widening concern for animal welfare (Dawkins, 1980) has facilitated the reception for cognitive analyses.

Carolyn Ristau (1986) has been another leading proponent of the usefulness of cognitive approaches to motivational analysis. She argues that for a variety of topics in animal behaviour such as honeybee dances, courtship in baboons, use of language by apes, and concept learning, an intentional analysis may be appropriate. Reminiscent of Bertrand Russell's theory of types, this research strategy discriminates different orders or levels of statements. Zero order, which is strictly behavioural, is of the form 'A did X'; first order is 'A knows that B did X'; second order is 'A believes that B knows that C did X'; and so forth. Working with the piping plover (*Charadrius melodius*), Ristau has attempted to reject the null hypothesis that zero order statements are sufficient to describe her observations. Her intentional analysis has focused on nest guarding activities which serve to distract potential predators. These activities include injury feigning (broken wing displays) and false brooding, in which a bird sits at some location other than the nest as though brooding eggs, and many even fluff their wings. Ristau argues that an endangered bird should monitor the behaviour of the predator for its movements and attention towards the nest; that it should react differently depending on this behaviour; and that if its eggs have been destroyed it should cease displaying. The obvious criticism here is that these predictions can also be derived by considering the consequences of such activities, whether phylogenetically through the selection of releasing mechanisms sensitive to eggs and intruders, or ontogenetically in terms of habituation or reinforcement of responses effective in the variety of situations involving the distraction of predators. In particular, Ristau and like-minded cognivitists refuse to specify what data would distinguish their interpretation from more traditional and parsimonious ones. This refusal flies in the face of the well established principle in science, most fully elaborated by Karl Popper (e.g. 1962), that hypotheses achieve standing only to the extent that observations can be specified which would falsify them. As William James (1902) put it, 'every difference must *make* a difference', and this is what cognitive ethology fails to do.

The problems associated with the natural history of mind have long proved a quagmire to the careless. Some progress has been made in recent years. For instance, Nicholas Humphrey (1983) has considered the evolution of consciousness. He argues that there is a significant selective advantage in the ability to model reality, and goes on to point out that, for highly social animals, the most important portion of reality to be modelled is the social group. Since such a group is composed of complex individuals, it is difficult to model, but the goal is achieved by use of the animal's own motives and behaviour as a source of analogy. In brief, the animal introspects.

The need for the animal to include itself in its model leads to self-recognition. Thus for Humphrey, consciousness has been shaped by natural selection to be a valid description of motivational mechanisms. Such speculation provides a provisional account of the distribution of consciousness across species.

However, most publications dealing with the mind are on a metaphysical par with the Franciscan regard for animal siblings, and lack any sensible basis. The writings of cognitive apologists are filled with conditional hints that such-and-such a notion may be needed for explanation, but are conspicuously devoid of any firm cases. More than ever, Lloyd Morgan's (1909) canon that 'in no case may we interpret an action as the outcome of a higher psychical faculty, if it can be interpreted as the outcome of the exercise of one which stands lower in the psychological scale' is needed to ensure the appropriate explanatory rigour.

The controversy between behaviouristic and cognitive ethology as presented by Griffin and Ristau involves some very fundamental issues. Those sympathetic to the latter regard the inclusion of a mental vocabulary as necessary for a full understanding of behaviour. By contrast, behaviourists view such a step as opening a Pandora's box of metaphysical plagues, to the detriment of the enterprise of accounting for data in the most parsimonious, competent language. Although only the most radical behaviourists continue to exclude intervening variables and clearly specified models, the economy of the theoretical baggage is still carefully regarded. While different terms are used at different levels of analysis, science is nevertheless unified by the fundamental feature of the mutual translatability of these terms. The central defect of mentalism is the admitted non-translatability of intentional terms into behavioural terms (a feature labelled referential opacity by Dennett, 1983). There can be no historical doubt that behaviourism has advanced ethology as a science, whereas the methods advocated by cognitivists have yet to prove their worth. Until mental concepts are clarified and their need justified by convincing data, cognitive ethology is no advance over the anecdotalism and anthropomorphism which characterized interest in animal behaviour a century ago, and thus should be eschewed.

# 3
# *Physiology of motivation*

---

Researchers into motivation often wish to pursue the behavioural phenomena in which they are interested to their component elements and processes. Reductionism is the research approach whereby systems of interest are broken into their component elements for purposes of generating explanations in terms of these elements, and was described by Nagel (1961) as follows: 'Reduction ... is the explanation of a theory or a set of experimental laws established in one area of inquiry, by a theory usually though not invariably formulated for some other domain'. In its extreme form, as developed in logical positivism, reductionism aims at descriptions of biological phenomena entirely in the language of physics. Reductionism thus inspires those workers seeking the nervous and hormonal mechanisms underlying behaviour. In the case of behavioural causation, the reduction of motivation to physiology is sought. Conflicting with reductionism is the attitude of emergentism, i.e. that at any specified level of organization there are phenomena which emerge unpredictably from the interactions of the underlying components: 'An emergent thing ... is one possessing properties that none of its components possess' (Bunge, 1977). Churchland (1986) presents an excellent review of why reductionism is a philosophically sound strategy for behavioural science.

While focusing on the behaviour of animals, ethologists have long appreciated the importance of physiology. For instance, aware that physiological condition affects motivational responsiveness to stimuli, Lorenz has cited Goethe's 'With this drink inside you, you will soon see Helen in every woman'. Beyond motivation, physiological researchers are interested in areas such as sensation and motor action, including orientation; and learning, involving the quest for that Holy Grail, the engram, or physical basis of memory. The chief tools used in these studies have been ablation, stimulation (by chemicals, especially hormones and neurotransmitters, or electric currents), and recording of neural activity (of single cells and sets of neurones, in intact and isolated nervous systems), and subsequent comparison with behavioural observations. An important methodological feature of physiological investigations is the performance of

sham operations or injections on control animals in order to allow
for any possible effects of the operation or injection itself. For instance, in a lesion study, control animals are cut open but no lesion is
made; while in an injection study, saline solution with a concentration
equal to the blood, or a vehicle without the drug or hormone, is
injected in a volume equal to that for the experimental animals.
Some studies include both unoperated and sham-operated control
groups.

The physiology of motivation has a large literature including the
review by Whalen and Simon (1984) and major tomes edited by
Lissak and Molnar (1982), Pfaff (1982), and Satinoff and Teitelbaum
(1983). Many of the classical topics in this area have been reviewed
by Gallistel (1980). These include the reflex, integrative action, and
final common path of the nervous system from C.S. Sherrington
(1857–1952), the oscillators from Erich von Holst (1908–1962)
(Holst, 1973), and the neural hierarchies from Paul Weiss (1898–   ),
Stellar and Stellar (1985) have surveyed the physiological hardware
underlying motivation and reward in rats.

A major department within the physiology of motivation is the
pharmacology of performance, i.e. the study of the effects of drugs
on activities. Much of this work involves the use of agonists and
antagonists, drugs which mimic or oppose the operation of neurotransmitters. For instance, there are motivational aspects to the
opioids, endogenous opium-like peptides synthesized in vertebrates.
Opioids include enkephalins and endorphins, which are produced
by the limbic system and pituitary gland respectively, as well as
by other sites in the body. Their study is facilitated through the
use of agonists and antagonists such as morphine and naloxone,
respectively. Opioids and their agonists serve as excellent positive
reinforcers (Stolerman, 1985). However, the evidence for the converse idea, that opioid antagonists should act as negative reinforcers
to condition escape and avoidance behaviour, is unclear.

From among numerous possible topics in the physiology of motivation, we shall consider in this Chapter five case studies which illustrate the diversity of exciting investigations currently in progress:
invertebrate systems, homeostasis, arousal and the reticular activating
system in cats, hunger and brain lesions in rats, and hormones and
sexual drives in canaries and lizards.

## 3.1  INVERTEBRATE SYSTEMS

For many physiological questions, the relatively simple nervous
systems and activity patterns of invertebrates stand as good candidates for investigation. One such candidate, the flatworm *Microsto-*

*ma*, was appreciated for its possibilities by the neuropsychologist Karl Lashley (1890–1958) who began his famous review on the 'Experimental analysis of instinctive behavior' (1938) with a summary of its diverse behaviour. Part of this behaviour includes the capture of hydras as a source of stinging cells for use in defence and prey capture. Lashley concluded that 'here, in the length of half a millimeter, are encompassed all of the major problems of dynamic psychology. There is a specific drive or appetite, satisfied only by a very indirect series of activities, with the satisfaction of the appetite dependent upon the concentration of nettles in the skin'. Lashley no doubt smiles from Avalon at the cellular analysis of behaviour so brilliantly carried out by Sidney Brenner (1974) and colleagues on roundworms. Other invertebrates have also been found to be fruitful research organisms, and many of these are reviewed in the volume edited by John Fentress (1976). In this section, 'the hungry fly' and sea hares will serve as exemplar invertebrate systems in the study of motivation.

## 'The hungry fly'

For Vincent Dethier, the hungry fly (1976), specifically the blowfly *Phormia regina*, has proven to be a rich system for studying feeding motivation. His experiments have uncovered the operative sensory and motor mechanisms involved, the effective external stimuli and internal control links, and details on such aspects as habituation and specific hungers in gravid females. With respect to the latter, such females deprived of proteins orient selectively to odours of putrefaction and ignore smells that attract them when hungry. (Specific hungers and diet selection have also been studied in vertebrates, especially rats (Ackroff *et al.*, 1986; Woodside and Milleline, 1987).) Similarly, a reproductively primed fly ready to copulate is influenced by pheromones, chemical signals emitted by conspecifics, but not by food. Gustatory stimuli both initiate feeding and inhibit locomotion, thus co-ordinating these antagonistic activities.

As well as specific hungers, central excitatory states comprise an aspect of Dethier's work which is of particular interest from a motivational viewpoint. The existence of such states in the central nervous system is inferred from the effects of food and water stimuli. These effects include an increase in excitation through repeated stimulation, a decay in this excitation over time which is dependent on stimulus intensity, and a decrease due to inhibitory stimuli. As an example of the tests used to measure such states, the assessment of the effects of stimulation with sucrose serves well. Flies which had been allowed to drink freely were pre-tested with a drop of water

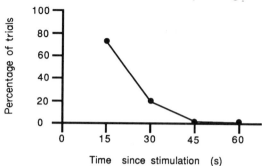

**Figure 3.1** Percentage of trials on which a blowfly extended its proboscis in response to water on the anterior labellar hair as a function of time since stimulation of a posterior hair with sucrose solution (from Dethier, 1976).

presented to an anterior labellar hair for five seconds. Such flies do not extend their proboscis. In contrast, presentation of a drop of sucrose solution to a posterior hair almost always did result in such extension. At intervals varying from 15 to 60 seconds after this sucrose stimulation, a drop of water was once again presented to the anterior hair, and extension of the proboscis was recorded. Individual trials on each fly were separated by intervals of two minutes. As Figure 3.1 shows, the percentage of trials on which the proboscis was extended at the second presentation of water decayed as the interval since sucrose stimulation increased. These results illuminate the dynamics of a central excitatory state.

As Dethier points out, care must be exercised to avoid modifying the state to be measured by the measurement procedure itself. This is a methodological point of general and great importance in motivational research. (It is an interesting analogue to Heisenberg's Uncertainly Principle in atomic physics, which asserts that both the position and velocity of a particle cannot be known. The usual appearance of this Principle in behavioural science is its invalid invocation in attempts to salvage free will in the face of determinism. But to deal with this topic would lead us far afield.)

While the onset of feeding is controlled by excitatory chemosensory input to the tarsi and proboscis, its termination is due to adaptation by these sensory receptors and especially to inhibition from stretch receptors in two locations (Bowdan and Dethier, 1986). Receptors belonging to the recurrent nerve act slowly and cumulatively in the foregut while similar receptors belonging to the abdominal nerve in the abdomen act immediately to shut off ingestion. The foregut is very active during feeding, and so its receptors are highly stimulated. The gradual action of these receptors prevents premature cessation of feeding, and this is probably achieved by the release of a neuromodulator into the central nervous system. This dual system of slow inhibition from the foregut and swift inhibition from the abdomen precisely regulate ingestion in the blowfly.

Along with information about endogenous oscillators producing behavioural rhythms, the concepts of specific hungers and central excitatory states enable explanations to be formulated for the variability in behaviour seen in blowflies beyond specific stimulus-bound reactions. Thus motivational analysis in this and other insects is far in advance of that for any vertebrate. Based on these extensive studies, Dethier has been able to provide a comprehensive under-standing of the behavioural processes of this animal (and, inciden-tally, written in prose of grace and clarity that ranks with the finest in the literature of contemporary biology as hallmarked by Stephen Jay Gould, Peter Medawar, and Lewis Thomas).

Behavioural physiologists, ethologists, and psychologists who, like Dethier, have investigated motivational phenomena carefully, have also frequently been led to postulate excitatory and inhibitory states in the central nervous systems of their study organisms. In his influential text on physiological psychology, C.T. Morgan (1943) argued that since 'the nervous system is the locus of inter-gration into which motivating factors pour and from which patterns of motivated behaviour emerge ... we might as well recognize such neural integrative activity and give it a name, the central motive state'. This concept has been found useful by many researchers since its enunciation by Morgan.

## Sea hares

The behaviour of sea hares, opisthobranch gastropods of the genus *Aplysia*, has been the object of much research by Eric Kandel (1979) and his colleagues. In terms of motivation, the feeding behaviour of *Aplysia* reflects a hunger–satiation continuum. Chemical stimuli from seaweed excite feeding while bulk in the gut inhibits it. Animals can be sensitized to make feeding responses by repeated presentations of food, and are then less likely to show defensive acts such as siphon withdrawal and inking. (Sensitization was discussed in relation to learning and habituation in section 2.3.) The overall arousal of a sea hare varies from a balled posture in which a stationary animal maintains a round shape with contracted tentacles and neck, through successively more extended postures to locomotion. Sensitization comes about from external stimuli which are neuronally mediated. Arousal is a result of both external and internal stimuli involving both nerves and hormones.

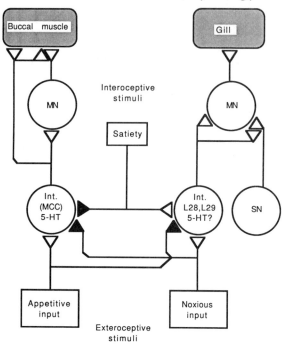

**Figure 3.2** Neuronal model of sensitization, arousal, and motivational state for feeding and defence in *Aplysia*. MN represents motor neurones; SN: sensitization neurones; Int: interneuronal neurones; and open and closed triangles excitatory and inhibitory connections, respectively (from Kandel, 1979).

Kandel provides a speculative model linking sensitization, arousal and motivation in feeding (involving the buccal muscle) and defence (involving gill withdrawal) in *Aplysia* (Figure 3.2). Appetitive and noxious inputs have antagonistic effects while satiety differentially influences the motor systems of the buccal muscle and gill. The motivational state of the interneuronal centres is influenced by both external and internal cues. This state determines the sequence of goal-directed activities, such as feeding, and mediates reinforcement, so that food influences the behaviour of hungry animals only. The model thus provides a physiological account for the motivated behaviour of *Aplysia*.

A similar physiological model of motivation has been put forth for *Pleurobranchaea*, a carnivorous cousin of *Aplysia*, by Jack Davis and co-workers (e.g. Kovac and Davis, 1980a; Davis, 1985) (Figure 3.3). The behavioural repertoire of this gastropod includes feeding, mating, egg laying, postural righting, local withdrawal, and escape

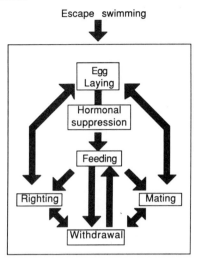

**Figure 3.3** Behavioural hierarchy of *Pleurobranchaea* (from Kovac and Davis, 1980a).

from predators by swimming. The behavioural organization of this animal includes both hierarchical (unidirectional) and reciprocal (bidirectional) elements. As expected, avoiding predators takes top priority. Feeding usually takes precedence over righting, withdrawal, and mating, as shown by a variety of observations. Animals maintain an inverted posture when liquefied food consisting of raw squid is squirted onto the oral veil. Touching the veil normally produces withdrawal, but if food is also presented, withdrawal is inhibited. Conversely, feeding is depressed by withdrawal. Both of these reciprocal effects are graded in response to the intensities of the stimuli. The concentration of liquefied food necessary to elicit feeding provides an operational definition of hunger level.

In *Pleurobranchaea* even mates are abandoned in favour of food, presumably because the latter is rarer than the former in nature. Indeed, in the herbivorous garden snail, *Helix aspersa*, feeding does not dominate mating in the behavioural hierarchy (Everett *et at.*, 1982). Returning to *Pleurobranchaea*, feeding is subordinated to egg laying via a polypeptide hormone stored in identifiable cells of the pedal ganglion, and this prevents ingestion of offspring. The motivational flexibility of this behavioural hierarchy is demonstrated with satiated individuals in which food inhibits righting, via chemosensation, but does not inhibit withdrawal. The behavioural organization can also be altered through learning. Pairing food with response contingent shock leads to avoidance of the food.

The avoidance is food specific, with responsiveness to homogenate of sea anemone remaining unaffected when shock is paired with squid homogenate.

At the level of the nervous system, the circuits underlying feeding and withdrawal have been explored (Kovac and Davis, 1980b; Croll and Davis, 1981; Davis, 1984). Feeding can be manipulated by nutritional history, and is initiated by 16 paracerebral command neurones innervating the buccal ganglion. These neurones also figure in generating the cyclic pattern of motor output which underlies feeding, and modulate this activity as a result of experience. In satiated animals, food stimuli do not excite but actually inhibit activity in these neurones, possibly due to input from oesophageal stretch receptors. Cyclic ingestion of food and egestion of unpalatable items involves the same central nervous oscillator and motor neurones, with shifts in the activity levels of the two elements.

The motor neurones of the withdrawal reflex are activated by sensory inputs directly and also by a command interneurone. The suppression of withdrawal by feeding is executed by central neurones without sensory feedback, as shown by activity in nervous systems isolated from such feedback. The discharge of a pair of identified neurones in the buccal ganglion of the feeding network is necessary and sufficient to inhibit withdrawal output. These neurones operate via corollary discharge; that is, they send a corollary, or copy, of the nervous output from the buccal ganglion to the brain during feeding. Subordinate activities are consequently inhibited by both central neural firing and sensory stimuli associated with the dominant activity. Finally, a linkage in the mechanisms of motivation and learning has been found by studying the isolated nervous systems of animals which had been conditioned with feeding stimuli. The results indicate that avoidance conditioning of feeding behaviour involves processes in cells presynaptic to the command neurones which underlie satiety effects.

Returning to *Aplysia*, a recent study has revealed the importance of gut distension and sensory stimulation as cues controlling satiety in feeding (Kuslansky *et al.*, 1987). Experiments in which the gut was filled with a non-nutritive gel via an oesophogeal cannula and the animal then permitted to feed indicated that satiation is determined by gut volume. In a similar vein, inflation of a balloon, either swallowed or cannularized through the body wall, also slowed the rate of feeding, and this effect was reversible through deflation. On a variety of behavioural measures, such as the latencies and amplitudes of bites, satiation due to food is similar to that due to inflation of a balloon. The rate of gut distension is also important, with fast-feeding animals requiring greater volumes to satiate. Cutting the oesophageal nerve produces enhanced feeding, just as the similar cutting of the recurrent nerve did in the blowflies studied by Dethier. Cues from the gut are accompanied by sensory stimuli in influencing satiety, as shown by the observation that prolonged exposure to food decreases the subsequent intake of food.

Thus gastropods are excellent test systems for the study of motivation, related behavioural phenomena such as learning, and the physiological substrates of these processes. In particular, they are valuable for providing instances of how choice behaviour is achieved physiologically through mechanisms such as interactions of neural centres via efference copies to sensory pathways and corollary discharges which inhibit activity. It is the treatment of such findings in the context of cybernetic theory which is currently providing the basis for a ·comprehensive understanding of motivated behaviour. Additionally, the neural circuits revealed in these studies indicate how motivational systems converge on the final common path via command neurones which initiate and drive organized activity.

## 3.2 HOMEOSTASIS

The concept of homeostasis as the descriptor of the steady state of healthy animals has loomed large in the physiological and ethological analysis of motivation. Homeostasis has historical roots in the classical studies of Claude Bernard (1813–1878) who described mechanisms which stabilized the *milieu interieur* of the cells of the body (e.g. 1927); Walter Cannon (1871–1945) who coined the term and had a particular interest in the physiology of emotions (e.g. 1919), and Curt Richter (1976) whose psychobiological work centered around homeostasis. A common distinction has been between homeostatic (drive-induced) systems such as breathing, feeding, drinking, and temperature regulation (all activities which maintain a steady internal physiological state) and non-homeostatic systems, such as aggression, reproduction, exploration, and play, which are highly cued by external incentive stimuli. However, this distinction is of dubious validity, as we shall see.

Although homeostasis is a functional term, there has developed through the application of cybernetics and systems theory almost an identity between homeostasis and mechanisms involving setpoints and feedbacks. (Cybernetics, systems theory, and feedback were discussed at the beginning of Chapter 2.) Nevertheless, homeostasis can be achieved by other mechanisms such as feedforward, in which an equilibrium is maintained by an action occurring before the equilibrium is upset, and mechanisms not involving feedback at all. Neither does the presence or absence of feedback parallel internal versus external control.

The basis for these conclusions has been provided by several contributors, such as Jerry Hogan, to the volume edited by Toates and Halliday (1980). Data on feeding and drinking, for instance, are not usefully interpreted in terms of a dichotomy between homeostatic and non-homeostatic processes. While some aspects of feeding and drinking are homeostasis, others, such as gorging on highly palatable foods, are not. Homeostasis cannot in any unqualified way be equated with set-point mechanisms. The correspondence between physiological regulation and behavioural processes is not a perfect one, and not all activities can be linked to a physiological need. This has been shown, for instance, in experiments by Hogan and others on young chicks and rats, where feeding is homeostatic but feedback mechanisms develop only later. The experimental chicks lived on food, sand, or bare floors. At one and two days of age, chicks in the same nutritional state pecked at different rates while chicks in different states pecked similarly. By the fourth and fifth day, nutritional state and pecking were associated. In the case of the rat pups during the first two weeks after birth, nipple search and attachment were not affected by deprivation. Thus in both chicks and rats the development of a functional association between physiological need and ingestive behaviour takes some time.

Hogan also found that feeding and breeding activities in hens provide insights into the relation of homeostasis and motivation. During incubation, hens lose about 15% of their body weight and eat about 30% of their usual daily intake. Experimental manipulations such as food deprivation or making food available near the nest show that the same factors influence feeding and body weight in brooding and non-brooding hens. The decreased feeding observed in brooding hens can be interpreted in terms of altered set-points, but can also be viewed as the outcome of interactions between motivational systems of feeding and breeding. (Such interactions are discussed in section 2.5.)

Hogan goes on to show that a distinction of homeostasis versus non-homeostasis is not supported by results of studies on feeding and aggressive displays to a mirror in Siamese fighting fish, or on feeding and responses to obtain nest material in female mice. The fish were studied in an operant situation in which the reinforcement was food or the opportunity to display to a conspecific (Figure 3.4 (a)). With food, as the schedule of reinforcement was altered so that more responses were required for a reward, the number of responses increased so that total food intake was sustained. Regulation was not seen, however, after prefeeding. By contrast, when sight of a conspecific was the reinforcer, responding remained constant and so reinforcements fell. Further experiments showed that whether or not priming (in which external stimuli increase drive level, as

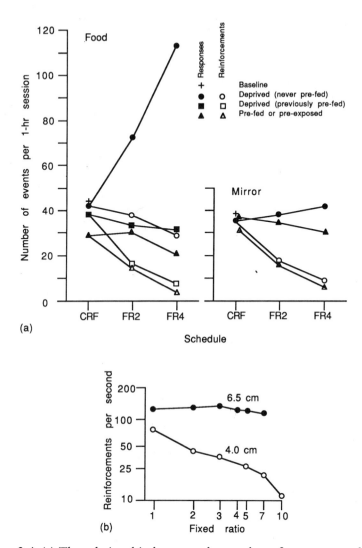

**Figure 3.4** (a) The relationship between the number of responses and reinforcements against the number of responses required for a reinforcement depends on the type of reinforcement (food or mirror) and recent history of exposure to this reinforcement in Siamese fighting fish. (CRF: continuous reinforcement; FR2: fixed ratio 2; FR4: fixed ratio 4.) (b) The same relationship depends on the distance between the responding disc and the dispenser of the reinforcer for nest material in female mice (from Hogan and Roper, 1978).

discussed in section 2.3) of display activity was obtained depended
on the experimental protocol. Swimming through a tunnel was not
primed by presentation of a mirror whereas runway performance
was enhanced. Similarly, prefed mice show little regulation when
making operant responses for food, and whether the acquisition of
food or nesting material is regulated depends on such features as the
proximity of the disc to be pressed and the dispenser of the reinforcer
(Figure 3.4 (b)). Thus experimental conditions determine whether a
system appears homeostatic, and priming can occur with both homeostatic systems such as ingestion and non-homeostatic ones such
as aggression.

Hogan's conclusion that 'it seems inappropriate to apply the
concept of homeostasis to the analysis of motivational processes'
reflects the importance of keeping causal and functional accounts
of behaviour separate. Causally, much work remains to be done
to investigate what mechanisms, involving feedback or otherwise,
underlie the regulation of activity through behavioural states. Systems theory promises to be instrumental in this work. Functionally,
these various drives are best appreciated by viewing an animal as
a co-ordinated system of components, all of which are needed for
maximizing genetic fitness.

## 3.3  AROUSAL AND THE RETICULAR ACTIVATING SYSTEM IN CATS

Understanding of the physiological basis of arousal, a phenomenon
closely linked with drive and attention, was advanced in 1949
when Moruzzi and Magoun described the reticular activating system
(RAS) in cats. Anatomically, the RAS consists of a diverse network within the central nervous system which ramifies in ascending
pathways from the brainstem to locations in the mid- and forebrain,
and back again in descending ones (Figure 3.5). The RAS is characterized by diffuse input and output connections, with ascending
cells typically responding to more than one sensory modality. The
electrical activity in the cortical regions of an awake and calm cat
typically exhibits synchrony. Stimulation of the RAS interrupts this
synchrony while lesions in the brainstem can produce animals who
remain in a deep sleep, much like students in an 8 a.m. lecture. Such
somnolence occurs after lesions of the medial regions of the reticular
activating system, but not after lesions of the lateral sensory pathways. Somnolent cats can be aroused by strong auditory and tactile
stimuli, and generally return to a normal cycle of wakefulness and
sleep after several days. Much research has been devoted to the
activity and function of the RAS, both in maintaining general alert-

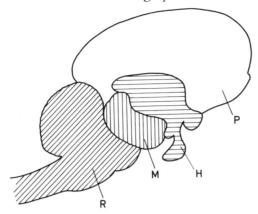

**Figure 3.5** Anatomy of the RAS in the rhombencephalon (R), mesencephalon (M), and prosencephalon (P) (including the hypothalamus (H)) of the cat in sagittal section.

ness and in influencing the operation of specific sensory systems such as vision, audition, chemosensation, and somatosensation.

Observations at the level of individual cells in the medial brain stem of cats show that the RAS is active in many behavioural contexts such as conditioning and habituation (Siegel, 1979). Habituation is reflected as a gradual decline in responsiveness to repeated stimuli due to changes in the sensitivity within the central nervous system to input from sensory receptors. The course of habituation can be followed in individual neurones of the RAS. Activity in the RAS is found in association with such motivating and emotional situations as anticipation of reinforcements, fear of electrical shocks, and agonistic behaviour. Anatomically, the distributions of cells involved in various activities overlap greatly, and any given cell is involved with several activities. These multiple activities are consonant with the general arousing property of the RAS. Anatomical, physiological, and phylogenetic evidence all suggest that neural discharge is related to the excitation of specific muscle groups, such as the eyes, tongue, or neck. This interpretation explains why the same cell can be active in conflicting behavioural contexts, such as anticipation and fear. Mediation of complex behaviour could be by other structures in the brainstem, smaller cells in the RAS, or as an emergent property of neural networks. Such an interpretation, Siegel points out, is consistent with the argument of the behavioural neurologist Roger Sperry that the nervous system is best understood as a mechanism for governing motor activity, which is the basis of overt behaviour.

Gross neural activity is the mass action of thousands of firings by individual cells as detected by recording electrodes. Often, such activity has a rhythmic pattern with a characteristic frequency, and is labelled by a Greek letter. Neuroscience has traditionally distinguished between phasic activity involving specifically patterned sensorimotor events, and tonic activity underlying motivational arousal. However, tonic activity is important in some sensory systems such as the somatosensory system. In terms of motivation, there is a correlation between tonic discharge rates in midbrain sites and observed behavioural arousal in cats. Bambridge and Gijsbers (1977) set out to examine the role of tonic activity in motivational processes during naturally occurring behaviour such as resting, play, eating, sleep, and paradoxical sleep (during which brain patterns resemble those of an awake animal). Multiunit recordings were made from various reticular areas as well as from two control sensory nuclei, the lateral geniculate nucleus and the inferior colliculus, involved in vision and audition respectively. Cats with chronically implanted electrodes were observed in an enclosure containing a tray, mat, food and milk.

The expected covariation of activity in reticular areas and behavioural arousal was found. Neural activity in a quietly sitting cat was 5–10% above that in slow-wave sleep, while in moderately to highly aroused cats, or cats in paradoxical sleep, activity rose 20–50%. The multiunit activity reflected not only general arousal but also rapid short-term behavioural changes such as reacting to novel stimuli. Cats quickly habituated to regular stimuli, displaying no behavioural or neural response. As a cat relaxes progressively from behavioural activity to sleep, there is a gradual decline in tonic levels of activity, interrupted only by novel stimuli and movements. This decline is greatest during the transition from drowsiness to sleep. By contrast, recordings by electroencephalogram on the surface of the cortex tend to emphasize differences between sleep and wakefulness but do not discriminate levels of arousal in an awake animal. The two sensory nuclei showed very different, sense-specific patterns of activity compared with that at the reticular sites.

The pattern of tonic activity in the RAS suggests control not by sensorimotor processes but rather by some cerebral co-ordination. The activity is altered by significant events such as the approach of the experimenter but not by sensorimotor events associated with, say, stroking. Thus, tonic activity in the RAS is closely associated with behavioural arousal. The central nervous system is usefully viewed as a set of specific neural centres, linked by feedback loops, which generate different motivational states. For both sensorimotor and motivational factors, increases in tonic activity are associated with increases in overt function. Finally, as discussed in section 2.6, these observations of discrete behavioural changes underlain by continuous physiological trends make one (at least, this one) think of catastrophe theory. This theory could be used to specify a control space representing continuous sensorimotor and motivational factors which jointly determine discrete behavioural states of arousal and sleep.

Arousal is not solely the function of the RAS. Routtenberg (1968) has argued that the RAS is one of two mutually inhibitory arousal systems. The RAS affects the probability of motivated responses occurring by organizing them into sequences. A second, limbic-midbrain system operates primarily to mediate the effects of rein-forcement, controlling responses by processing incentive stimuli as well as such experimental inputs as electrical self-stimulation. The distinction of these two systems, which are anatomically close in some regions, resolves conflicting experimental observations such as the effects of lesions and drugs on sleep and wakefulness. For instance, it appears that some drugs, such as chlorpromazine, act on the limbic midbrain system, while others, such as the barbiturate pentobarbital, affect the RAS. A schema summarizing the anatomi-cal and physiological relations of these two behavioural systems is shown in Figure 3.6, indicating their mutual antagonism. Activity in the RAS is enhanced by the posterolateral hypothalamus but dampened by the lateral septal area. Thus, hypothalamic reward is exciting while septal reward is quieting.

A variety of effects can be understood with this schema. For in-stance, escape and avoidance responses are facilitated by concurrent rewarding hypothalamic stimulation which augments activity in the dorsal midbrain. Similarly, dorsal midbrain stimulation inhibits self-stimulation of the hypothalamus. The schema suggests that dorsal midbrain stimulation is aversive because it suppresses the septal system. Further, it indicates that hypothalamic self-stimulation, while increasing activity in the septal system, more importantly also enhances that in the hypothalamic system. In turn, this en-hancement adds to the ongoing activity in the dorsal midbrain. The overall activity is aversive and so self-stimulation is blocked. The rage and aggressiveness seen after septal lesions can also be understood in terms of the release of the RAS from septal inhibition. With the septal inhibition removed, the RAS can produce aggressive outbursts. The limbic system can dominate the RAS via the hippo-campal theta rhythm (with a frequency of 4–7 Hz), which thus plays a modulatory role. Neuropharmacologically, pentobarbital (a seda-tive hypnotic) and chlorpromazine (a tranquillizer) both depress acti-vities but through different pathways involving the RAS and limbic

Physiology of motivation

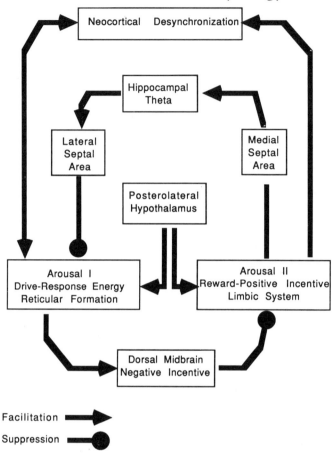

**Figure 3.6** Model of the interactions of the reticular activating system and the limbic-midbrain arousal system (from Routtenberg, 1968).

system, respectively. Finally, the schema also incorporates the close linkage between motivation and learning. Specifically, the suppression of the RAS by the limbic system via the septal route reflects the reduction of a drive producing a response by the effects of reinforcement. Routtenberg's schema suggests a neurological basis to several major motivational topics, and indicates the complex role of the RAS in these functions.

For those interested in the neural basis of choice and action in vertebrates, the relations of the RAS and cerebral cortex to behaviour are of central importance. As Vanderwolf (1983) points out, with emphasis on experimental work in rats, establishing these relations must be done carefully. It is trivially obvious that electromyograms and behaviour are correlated, and studies of the nervous system and behaviour should not be likewise confounded. It is necessary to make observations both while behaviour is constant and while it varies and, as the work with cats above indicates, different sections of the nervous system have very different functions. Activation in the neocortex varies with the area studied while that in the hippocampus is related to concurrent motor activity, with two patterns to be found. The first pattern is the nearly sinusoidal, slow theta rhythm associated with voluntary and appetitive acts, while the second pattern is faster and more irregular, occurring in connection with reflexive and consummatory responses (Figure 3.7). These two patterns are differentially affected by pharmacological agents, and the corresponding set of acts obviously parallel distinctions made by

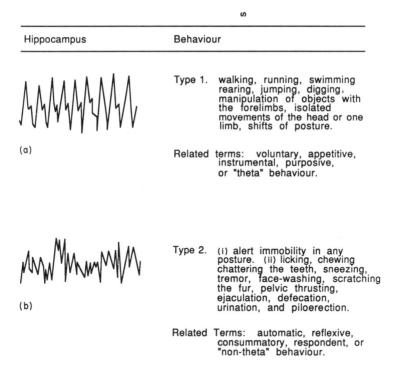

| Hippocampus | Behaviour |
| --- | --- |
| (a) | Type 1. walking, running, swimming rearing, jumping, digging, manipulation of objects with the forelimbs, isolated movements of the head or one limb, shifts of posture.<br><br>Related terms: voluntary, appetitive, instrumental, purposive, or "theta" behaviour. |
| (b) | Type 2. (i) alert immobility in any posture. (ii) licking, chewing chattering the teeth, sneezing, tremor, face-washing, scratching the fur, pelvic thrusting, ejaculation, defecation, urination, and piloerection.<br><br>Related Terms: automatic, reflexive, consummatory, respondent, or "non-theta" behaviour. |

**Figure 3.7** Two patterns of hippocampal activity associated with arousal: (a) slow, nearly sinusoidal; and (b) fast, irregular pattern (from Vanderwolf, 1983).

ethologists such as Craig (appetitive and consummatory responses) and psychologists such as Skinner (operants and respondents) in other contexts. The neocortex, too, shows two patterns of activation, and this finding helps to resolve earlier contradictions in the physiological literature since there are separate behavioural manifestations of each type of activation.

Further in the relation of these patterns to arousal, it turns out that neither pattern determines sleep. Decorticate animals maintain cycles of wakefulness and sleep, and both patterns occur during paradoxical sleep. Such animals are capable of many activities, although complex behaviour is disrupted, and not just because of sensorimotor impairment. The effect of decortication is to damage nodes, high in a hierarchy of control, which are important for selecting appropriate responses. The co-ordination for these responses is organized by lower nodes.

Jointly, these investigations reveal some of the intricate operations of the RAS in the cat as this elegant animal moves through its spectrum of arousal from intense concentration on a mouse through contented purring to blissful slumber.

## 3.4 HUNGER AND BRAIN LESIONS IN RATS

Few parts of the rat brain have remained inviolate to lesioning, chemical injection, and electrical stimulation from either the experimenter or the owner. These latter stimulation experiments (well reviewed by Halperin and Pfaff, 1982, and Stellar and Stellar, 1985) have helped illuminate general processes in motivation. For instance, the separation of non-associative, motivational effects of priming stimulation from associative effects during performance is an interesting comparative parallel to the neural distinctness of satiety and conditioning phenomena in *Pleurobranchaea* discussed in section 3.1. Thus investigations in both mammals and molluscs reveal one group of motivational phenomena involving arousal, priming, and drive level, and a second group involving association of stimuli and responses through conditioning. This section will focus on findings from lesion research, which is of interest because lesions produce motivational deficits, i.e. decrements in responsiveness independent of sensorimotor impairments. We shall examine the following:

1. the behavioural effects of lesions in one specific area of the brain (the hypothalamus);
2. an account of these effects in terms of ingestive reflexes;
3. experimental work focusing on sham feeding and the metabolic processes of insulin secretion and energy flow.

## Hypothalamic lesions

An overview of the use of lesions in studying feeding in rats is presented by Norgren and Grill (1982). As predominantly nocturnal animals, rats feed more frequently during the dark than during the light. Lesions in the ventromedial hypothalamus (VMH) have behavioural effects such as finickiness and hyperphagia as well as metabolic effects including obesity. Finickiness is seen as preference for highly palatable foods and aversion to adulterated foods, and this effect is greater in heavier animals. There is thus a link between finickiness and metabolic state, and this is further illustrated by regulation of food intake after experimental gastric loading. Even when limited to small meals, lesioned rats still become obese. Although lesioned animals eat heavily, they perform increasingly poorly on operant tasks to acquire food as they gain weight. When obesity is prevented by controlling available food, there is a relative increase in fatty tissue, and the polypeptide hormone insulin is involved in this change. Although it has been proposed that VMH lesions damage a satiety centre which normally inhibits the feeding centre in the lateral hypothalamus, the actual mode of operation is unknown. Indeed, of the several behavioural and physiological explanations put forth, most have proven to be inadequate for all the available data. It is known that endocrinal factors are important, as discussed below, and that the VMH receives input from the viscera and contains neurones whose rate of glucose metabolism determines firing.

## Ingestive reflexes

Terry Powley (1977) has postulated that VMH lesions lead to an exaggeration of the cephalic phase reflexes of ingestion which are dependent on sensory aspects of food (Figure 3.8). This leads to finickiness and, via visceral responses, to metabolic changes, overeating, and weight gain. In each of its cephalic, gastric, and intestinal phases, digestion involves autonomic and endocrinal reflexes. The cephalic phase reflexes arise in the head and oropharyngeal area and travel to the gastrointestinal tract and viscera where they control secretion, absorption, and motility. Sensory cues, especially taste and odour from food, activate these reflexes which facilitate both ingestion and digestion. Cephalic reflexes also include the release of various compounds, including digestive enzymes and hormones, and rejection responses such as gagging and vomiting. Evidence for such reflexes is gathered through experimental techniques such as sham feeding in which the oesophagus is brought to the exterior and so food is swallowed but not passed into the stomach. The importance

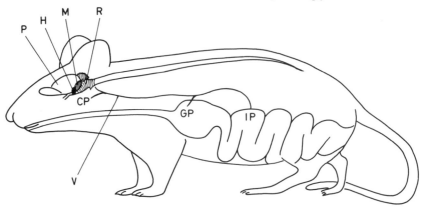

**Figure 3.8** Hypothalamus and contribution of the vagus (tenth cranial nerve) to digestive reflexes in the rat. The hypothalamus (H) is a lower part of the prosencephalon (P) (cf Figure 3.5). R, rhombencephalon; M, mesencephalon. Vagal innervation (V) for digestion has cephalic, gastric, and intestinal phases (CP, GP, and IP).

of these reflexes in humans has been shown in individuals with stomach fistulae. Food placed directly into the stomach fails to satisfy appetite, which can be achieved only by the tasting and chewing of morsels. In his hypothesis Powley assumes that beyond the changes in the cephalic reflexes due to VMH lesions, other homeostatic mechanisms involving energy metabolism are operating normally.

Many of the experimental techniques used in this area were developed by Pavlov in the course of his study of conditioned responses. Pavlov, who was concerned chiefly with the process of conditioning, generally used artificial stimuli such as tones and lights, and focused on salivation as the response. Weingarten and Powley (1981) have emphasized natural food cues, and have shown that classical conditioning of the cephalic phase of gastric acid secretion occurs in the rat. Experimental rats were trained to anticipate food in a particular situation, and their gastric acid secretion, measured via a cannula system, was compared with that of control animals. This anticipatory response for food was rapidly brought under stimulus control. The complexity of the processes responsible for such conditioning keep this area of research controversial.

The cephalic phase hypothesis can account for the diverse observations on the effects of VMH lesions by drawing on a well established phenomenon in ingestive physiology. An exaggeration of cephalic reflexes following VMH lesions would lead to increased ingestion of palatable foods as well as aversion of unpalatable ones. When food is remote, as in operant situations, the reflexes are not activated, and finickiness and laziness result. Hyperphagia and obesity occur because the reflexes maintain ingestion, and digestive metabolism shifts toward storage. By way of further support for the hypothesis, there are correlations between the cephalic reflexes and features of the VMH syndrome. For example, finickiness is paralleled by the ease with which the reflexes can be disrupted and conditioned. There are also disturbances of the reflexes by VMH lesions as measured by changes in the production of compounds such as gastric acid and insulin. Finally, cephalic reflexes, which often involve the vagus nerve as an afferent pathway, are modified by stimulation of the VMH.

Under the cephalic phase hypothesis, the function of the VMH is seen as modulating the amplitude of anticipatory digestive reflexes in relation to body weight and nutritional needs. Certain data suggest, but do not compel, that the hypothesis should be generalized to an autonomic hypothesis including the gastric and intestinal phases of digestive reflexes as well as the cephalic phase reflexes. Overall, the hypothesis clearly and plausibly links observations on feeding performance with autonomic events known to be important in motivated and emotional responses.

## Sham feeding

Sham feeding is a useful procedure because it enables an experimental separation of oropharyngeal stimulation from subsequent stimulation arising in the stomach, intestines, and postabsorptive sites. Control of normal and sham feeding can be achieved by the closing and opening of implanted gastric cannulae. Cox and Smith (1986) employed both sham feeding and transection of the vagus nerve (vagotomy) in the abdominal region in order to study the action of such stimulation. The results on the volume (in ml) of sweet milk consumed in a one-hour test by intact and VMH-lesioned rats under normal or sham feeding can be tabulated as follows:

|     |                | Feeding |      |
|-----|----------------|---------|------|
|     |                | Normal  | Sham |
|     | Intact brain   | 6       | 13   |
| Rat |                |         |      |
|     | VMH-lesioned   | 15      | 57   |

That is, rats with intact brains drank an average of 13 ml of sweet milk when sham feeding compared with 6 ml when feeding normally. Experimental rats, lesioned in the VMH, drank 57 and 15 ml respectively. Thus the lesioned rats, which normally eat more than control animals, showed a much greater increase in consumption under sham feeding. Vagotomy largely removed this over-responsiveness as shown in a second experiment for which the results were as follows

|  | Feeding | |
|---|---|---|
|  | Normal | Sham |
| Intact brain | 8 | 15 |
| VMH-lesioned | 16 | 47 |
| Vagotomized | 6 | 13 |
| VMH-lesioned and vagotomized | 9 | 24 |

Vagotomy had negligible effects on the behaviour of unlesioned rats, whether feeding normally (8 versus 6 ml) or under sham conditions (15 versus 13 ml). By contrast, lesioned rats with their vagus nerves intact or cut consumed 16 and 9 ml, respectively, when feeding normally, and 47 and 24 ml when sham feeding. These results thus support Powley's hypothesis that the VMH syndrome involves exaggeration of orosensory cues. Vagotomy largely, but not entirely, normalizes the responsiveness of lesioned rats. The exact mode by which the vagus achieves this normalization remains to be investigated, as does the contribution of other mechanisms such as secretion of glucocorticoid from the cortex of the adrenal glands.

## Metabolic processes

Whatever the role of glucocorticoid secretion, another endocrinal factor mediating the behavioural effects of VMH lesions is insulin, produced by the beta cells of the pancreas. The importance of insulin secretion has been shown both by experimental work on the sensitivity of these cells to transmitters of the autonomic nervous system, such as norepinephrine, and by computer simulation of a systems model including nervous, endocrinal, energetic, and rhythmic components (Campfield et al., 1982). VMH-lesioned rats have high circulating levels of insulin due to increased parasympathetic and, especially, decreased sympathetic nervous activity. Pancreatic cells from lesioned rats demonstrate enhanced sensitivity to norepinephrine. Thus insulin is involved in the energetic dynamics producing obesity in lesioned rats, as has also been shown in a recent physiological investigation in which chemically induced lesions resulted in changes in both body weight and insulin activity (Berthoud and Powley, 1985). The simulation model of Campfield and co-workers accurately predicted both physiological and behav-

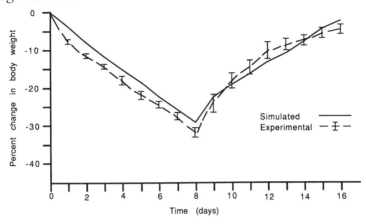

**Figure 3.9** Comparison of simulated and observed changes in body weight over 8-day periods of fasting and *ad libitum* feeding in VMH-lesioned rats (from Campfield *et al.*, 1982).

ioural aspects of the VMH–obesity syndrome, including insulin levels, feeding patterns, and body weight (Figure 3.9). Of several conflicting hypotheses investigated by using this model, the only satisfactory one incorporated insulin secretion being increased by nervous activity; elevated absorption of food across the gut during the light phase; and attenuation in the circadian rhythm of feeding thresholds and resting metabolism. This research shows the value of interfacing relevant experimentation and quantitative modelling.

The relationship between hunger and VMH lesions has also been investigated by David Booth and co-workers (e.g. Duggan and Booth, 1986) who are testing the theory that feeding motivation is controlled by the flow of energy from absorption. Aware that rats have a circadian rhythm of feeding, eating more frequently at night than during the day, Duggan and Booth measured rates of gastric emptying in the early part of the light phase when these rates are lowest. Rats lesioned four hours earlier or one week earlier, and obese lesioned rats, emptied their stomachs more quickly than did sham-operated control rats. These results can be interpreted as indicating that the satiating effect of food is abbreviated in lesioned animals, and so overeating and obesity result. Other factors such as increased insulin production and fat mobilization may be of only secondary importance. One implication of these findings is that the slowing of gastric emptying may serve as a useful therapeutic measure in the treatment of obesity.

Jointly, these studies demonstrate the diverse roles of nervous, endocrinal, rhythmic, and behavioural factors in the VMH–obesity syndrome. The research of the workers in this field is exemplary of efforts investigating the physiology of motivated activity.

## 3.5 HORMONES AND SEXUAL DRIVES IN CANARIES AND LIZARDS

The motivating effects of hormones on animal behaviour have been well known since Berthold's (1849) experiments in which the behavioural effects of castration and transplantation of testes were demonstrated in roosters and capons. Beyond intensively studied mammals (for review see Feder, 1984), canaries and lizards provide excellent cases illustrating the links between hormones and sexual drives.

### Canaries

Numerous species of birds have provided useful data on the topic of hormones and behaviour. For instance, the interplay of sexual hormones (such as testosterone and oestradiol) and stimulation from the mate, eggs, and offspring leads to reproductive and parental behaviour in doves (Figure 3.10). The investigation of this behaviour, as pursued by Daniel Lehrman (1919–1972) and succeeding workers, has served as a paradigm for this area of research.

In many species of birds, singing is one of the frequent activities of sexually active males as they establish territories and advertise for mates. The neural and hormonal basis of song learning and production has been an object of interest to a variety of researchers (e.g. Margoliash and Konishi, 1985). In canaries (*Serinus canarius*) this basis has been scrutinized in several laboratories, especially that of Fernando Nottebohm and co-workers (e.g. Nottebohm, 1984; Nottebohm *et al.* 1986, 1987). Much has been discovered about the wiring of this system. On a comparative basis, it is worth noting that there is no evidence in this species for brain lateralization similar to that found in the speech centres of human brains. In young male canaries subsong begins at about six weeks after hatching. Subsong is followed by plastic song which includes features of adult song, such as loudness and phrasing, but which exhibits a variable syllabic structure. By sexual maturity, canaries have achieved full adult singing ability. Unlike many other species, canaries are capable of learning new songs in the breeding seasons of subsequent years, and the vocal repertoire is larger in the second year of life than in the first.

On a seasonal basis, a period of song instability in the summer and autumn leads to an incorporation of new syllables in the

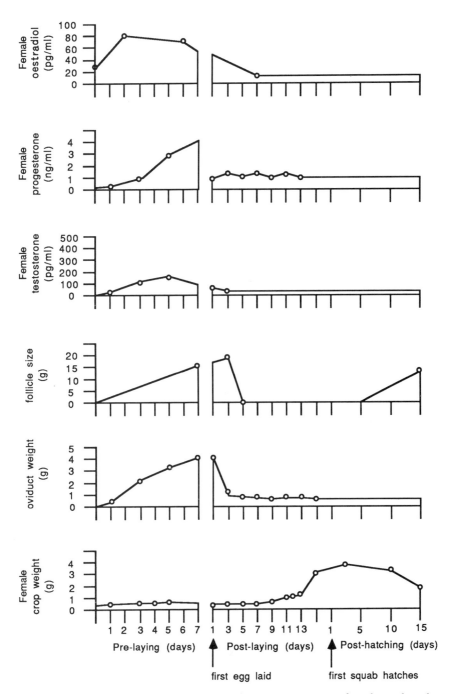

**Figure 3.10** Changes in hormones and target organs in female and male ring doves over the reproductive cycle (from Silver, 1978).

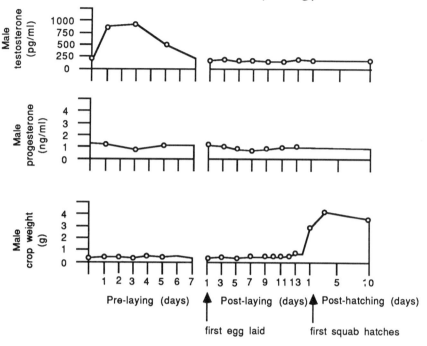

**Figure 3.10** (*contd*)

repertoire of the following breeding season during which song is stable. Concurrent with these seasonal changes in song repertoire are fluctuations in gonadal hormones, including testosterone, dihydrostestosterone, and oestradiol. Studies on canaries whose hormonal titres are measured while their singing is monitored reveal that the ontogeny of song is affected by all three of these hormones. In particular, the seasonal occurrence of recruitment of new types of syllables is correlated with high circulatory levels of testosterone. During the summer, testosterone levels are lower and song is less stable.

The developmental pattern of both singing and hormonal production is different in males and females (Weichel *et al.*, 1986). While levels of circulating testosterone increase in both sexes prior to the juvenile moult of plumage, males have higher levels of oestradiol overall. Further, the association of hormones and singing is not a tight one. For instance, the growth of stereotypy, or crystallization, of full song in males does not correlate with testosterone levels. Possibly a synergism between testosterone and oestradiol underlies the sexual differences in song development.

These changes in singing repertoire are mirrored in changes in the size of two forebrain nuclei involved in song control, the hyper-striatum ventrale pars caudale (HVC), which projects to the nucleus robustus archistriatalis (RA). This nucleus in turn links to the hypo-glossal nucleus in the brainstem from which originate the tracheosy-ringeal nerves innervating the muscles of the syrinx (Figure 3.11). With total brain size held constant, these nuclei are about two-thirds larger in spring as compared with autumn brains. Concurrently, the size of the testes and the circulating levels of testosterone are high in the spring and low in the autumn. Moreover, the volumes of these nuclei are similarly larger in ovariectomized females receiv-ing injections of testosterone (in doses producing circulating levels typical of these observed in adult males) compared with similar females receiving control injections of cholesterol but not testos-terone. The action of testosterone on the neurones of the nuclei is to induce growth in the dendrites, the cellular projections which re-ceive input from other neurones. Thus, it seems that the hormone-dependent seasonal swelling and shrinking of these nuclei underlies the learning and forgetting of songs. This conclusion is supported by other lines of evidence, such as the acquisition of songs during juvenile life, at which time the nuclei are growing rapidly. The plasticity of this causal chain involving testosterone, dendritic growth and re-sulting nuclear expansion, as well as updating of the song repertoire, is an exciting new dimension to the role of hormones influencing the nervous system and, through it, behaviour. As an oratorio chorister, I wonder . . .

Comparative information from zebra finches (*Taeniopygia guttata*) provides further insight into the sexual dimorphism of the auditory and song production system. This species does not learn new songs in adulthood, and as might be expected, the associated brain nu-

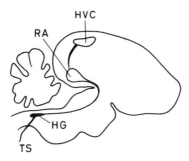

**Figure 3.11** Auditory and song system brain centres in male passerine birds. HG: hypoglossal nucleus; HVC, hyperstriatum ventrale pars caudale; RA, nucleus robustus archistriatalis; TS, tracheosyringeal motor nerve (from Williams, 1985).

clei are stable in volume once sexual maturity is reached. As in canaries, these nuclei as well as the syringeal muscles are larger in males as compared with females (Williams, 1985). The nuclei of females contain fewer and smaller cells, and are influenced by hormonal levels. As hatchlings, females can be masculinized by implantations of oestradiol. Like males, and unlike untreated females, such masculinized females show auditory responses in their tracheosyringeal motor nerves. These responses correlate directly with the sizes of the song nuclei. In normal females, cells of the HVC fail to excite those of the RA; however, treatment with oestradiol promotes this connection. These sex-specific aspects of song production may have consequences for song perception: since females lack the neuronal wiring which in males underlies song learning, they may be unable to extract as much information as males from songs.

More generally, comparisons among a number of species of songbirds show that the extent to which males and females differ in their song production is paralleled by various features of neuroanatomy such as nuclear volume. These correlations indicate exciting possibilities for exploring linkages among neural plasticity, flexibility in singing behaviour, and aspects of social ecology.

## Lizards

Another interesting case is that of hormones and sexual drive in lizards investigated by David Crews, Neil Greenberg, and colleagues (e.g. Greenberg and Maclean, 1978). *Anolis carolinensis* is a common iguanid throughout the south-eastern United States with an annual cycle of breeding and dormancy. Aggression and courtship are tightly linked and involve species-typical bobbing displays and extensions of a red dewlap, or throat skin flap (Crews, 1979). In females, the hormones oestrogen and progesterone synergistically promote sexual receptivity. After mating, females are no longer receptive. This loss in receptivity is a result of a neuroendocrinal reflex to sensory stimuli resulting from copulation, together with the effect of progesterone acting alone. Such a reflex is also known from mammals, but the causal associations between sexual behaviour and reproductive physiology are diverse across vertebrate animals (Crews, 1984). The loss in sexual receptivity in female *Anolis* is due to the action of the ovaries. Two lines of evidence support this conclusion. Ovariectomized females primed with oestrogen remate within hours. By contrast, sham-operated intact females who have mated and are then treated with oestrogen do not remate. Thus, progesterone both synchronizes sexual receptivity with ovulation, and terminates this receptivity.

Both gonadal and adrenal hormones also influence male social behaviour. Sexual behaviour is dependent on testosterone, and can be
reinstated in castrated males by injections or hormone-impregnated
silastic implants. In restricted environments, males form dominance
hierarchies, with dominant individuals displaying green skin while
subordinates are darker. Greenberg *et al.* (1984a) showed that subordinate lizards in such hierarchies are stressed, as reflected in the
level of the principal adrenal steroid, corticosterone. In an experiment, pairs of males were formed in which neither, one, or both
individuals were castrated. In mixed intact–castrate pairs, the intact
animal always became dominant. Corticosterone levels were similar
in dominant individuals, castrated subordinates, and isolated individuals, but elevated in intact subordinates (Figure 3.12). Intact
subordinates were also the only group to show significant skin
darkening. These results reveal the link between the social status of
an animal and its adrenal response. Furthermore, anatomical and
physiological research (Greenberg *et al.*, 1984b) has identified the
locations in the central nervous system at which these hormones

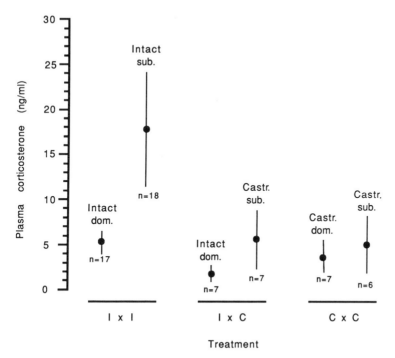

**Figure 3.12** Plasma corticosterone (mean ±2 sem) in socially paired lizards. I × I: intact–intact pair; I × C: intact–castrated pair; C × C: castrated–
castrated pair (from Greenberg *et al.*, 1984a)

have their effects. Thus a variety of hormones affect the motivation of lizards in their social interactions.

These studies on canaries and lizards are two exemplary cases of examinations of the physiological processes motivating courtship and mating. Unravelling motivational mechanisms is only one of many problems facing the investigation of reproductive behaviour. Other problems involve the roles of genetic mechanisms, sex determination and ratios, endocrinological and neurological pathways, recognition of sex and sexual arousal, the variegated influence of social cues, and adaptive aspects including sexual selection, alternative reproductive strategies, parental investment, and biological fitness. Reproductive behaviour is the complex result of the interaction of hormonal and neural mechanisms operating in a social and ecological context. Studies focusing on such motivational mechanisms must heed this context, just as in the case of feeding discussed in section 4.4. Finally, as of yet relatively few species have been studied in detail. It is therefore necessary to extend the comparative basis of available observations in order to scrutinize the validity of current generalizations as differences among species are encountered.

## 3.6  CONCLUSIONS

What are the prospects for the physiological analysis of motivation? Davis (1985) speculates on the possibility of extending his findings on the behavioural hierarchy of *Pleurobranchaea* to other species, including vertebrates with their vast nervous systems. Certainly such findings provide insight into mechanisms which may be elaborated in more advanced groups. But will the reductionist programme eliminate motivation as a behavioural topic? Some authoritative voices reply in the affirmative. For instance, Dethier (1976) compares the spontaneity and flexibility of his blowflies with that of brain-lesioned rats and finds no fundamental differences in their motivated activity. He proceeds to question the loose use of motivational terms by many investigators of vertebrate behaviour. After reviewing the outcomes of a variety of tests, Dethier concludes that 'none of the measures described ... support the need for a concept of motivated behaviour as usually defined because all are amenable to explanation in terms of known physiological processes as far as the fly is concerned, and probably also with respect to the rat'. He has gone on to argue that 'the fuzziness, the elusiveness, and the controversy surrounding the concept of motivation suggest that the idea has not only outlived its usefulness as an analytical scaffolding but has become an impediment to our understanding of any behavior that it purports to explain' (1982). Dethier's fellow entomologist, John Kennedy (1987), shares Dethier's discontent with the concept of motivation in his chapter of Dethier's *Festschrift*. More forcefully, Peter Morgane (1975), extending Frank Beach's complaint that motivation is the 'phlogiston of psychology', has asserted 'Does not "behaviour" have its essence in the underlying dendritic geometry, nerve impulse activity, and brain chemistry? What is the magic of the word "motivation" in coming to terms with the brain itself? We should not continue attempts to perpetuate a false dichotomy between behaviour and underlying physiological mechanisms, where "behaviour" becomes some sort of epiphenomenon divorced from underlying processes beyond the reach of physiological and biochemical approaches'.

These viewpoints notwithstanding, both physiological and behavioural approaches to motivation have legitimate uses at their respective levels. This dominant viewpoint is widely held by both ethologists and physiological psychologists such as the Stellars. In Richard Dawkins' felicitous phrase, we do not attempt to read a book with a microscope. Davis (1985) concludes that 'the concept of motivation furnishes a key unifying principle for diverse forms of behavioral plasticity'. In the terms of the computer analogy introduced in the previous chapter, behavioural physiologists seek hardware, or wiring, accounts of motivation while ethologists are interested in software, or programming, descriptions of the activities of animals.

# 4
# *Ecology of motivation*

The areas of greatest interest in biology change over decades, and most recently research in two regions, molecular biology and behavioural ecology, has been the most fervent. Indeed, the more messianic workers in these pursuits vainly attempt to make other biologists feel *déclassé*. The advances in behavioural ecology have been under the umbrella of adaptationism, which is the argument (truism?) that all major features of organisms for which there is selection and heritable variation, including their behaviour, will be optimized over generations. However, there are a number of reasons why behaviour may not be optimal. For example, the optima may shift seasonally or over longer periods of time. Phylogenetic, ontogenetic, or sensorimotor constraints on behavioural flexibility may limit the responses of animals. Additionally, optimization of overall fitness requires trade-offs among different activities such as feeding and reproduction.

The application of these Darwinian principles through specific theories dealing with such topics as optimal foraging, mating systems, and parental investment raises the need for a reconciliation of our accounts of proximate and ultimate mechanisms. Attempts (e.g. McNamara and Houston, 1986) at such a reconciliation centre around the search for a common currency in which to measure the costs and benefits of actions. The goal is to account for behavioural dynamics as these contribute to biological fitness, which is notoriously difficult to measure. Concurrent with the search for a common currency by functionally oriented workers, motivational researchers such as the Colliers (Collier, 1980; Collier and Rovee-Collier, 1983) are increasingly emphasizing the need to attend to the ecological context of, say, reinforcement. Thus, while behavioural ecologists are realizing the importance of understanding the causal basis of functioning activities, investigators of mechanism are paying more attention to the adaptive value of behaviour.

While adaptationism is no panacea for biological puzzles, it does often suggest solutions – too many solutions, critics maintain! – to diverse problems, including motivational ones. For instance, Hogan and Roper (1978) use the example of the lack of response in animals to carbon monoxide to illustrate the imperfect match between physiological need and motivational drive. An adaptationist (certainly this adaptationist) is quick to point out that animals do not usually encounter appreciable concentrations of carbon monoxide and therefore there has been no selection for a response. More generally, adaptationism provides a framework for evaluating behavioural evolution by examining those cases where the phylogenetic correlation of physiological needs and motivational drives is high.

Case studies in this chapter will deal with optimal behaviour; matching and maximizing; individual differences; hunger and foraging; and the signalling of intentions.

## 4.1  OPTIMAL BEHAVIOUR

Beyond the proximate mechanisms for behavioural sequences discussed in the previous chapter, David McFarland and co-workers have suggested a functional approach involving optimization techniques. The overall strategy has been to try to identify functions incorporating internal deficits and external availabilities which animals optimize through their behaviour. Specifically, in one well researched paradigm, the question of interest has been how a hungry and thirsty dove will reduce its deficits. McFarland envisages such a bird as occupying a position in a casual factor space with axes representing motivational factors, in this case hunger and thirst. The problem in this state space model is to account for the trajectory which the bird follows in this space as it reduces its needs.

It is assumed that the dove is maximizing its fitness through some function of its needs and activities (Sibly and McFarland, 1976). In general, it is assumed that the cost of a need in terms of fitness rises faster than linearly, say as a squared function of need. That is, the cost of a deficit of two aliquots of water is more than twice that of one. Once the overall cost function is specified, established mathematical techniques for optimization can be employed (providing you can remember how to spell either Hamiltonian or Pontryagin) to calculate when the bird should eat and drink. It is postulated that this choice is determined by a comparison of the product of deficit and incentive for food with that for water. Deficit is operationally defined as the quantity of food or water that the dove ingests until satiated, and incentive is the maximum rate at which food could be acquired in the operant situation. This definition is in the tradition which identifies behavioural drives with physiological needs.

The set of points in state space for which these products are equal is termed the dominance boundary. This boundary, whose slope is the ratio of the incentives, divides the space into regions in which

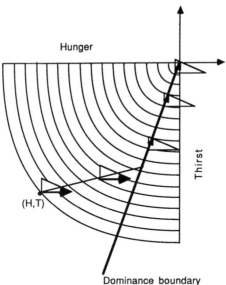

**Figure 4.1** An optimal trajectory for a bird beginning at position (H,T) in causal factor space spanned by axes representing aspects of hunger and thirst (from Sibly and McFarland, 1976).

either drinking or eating tendency is greater. A possible dominance boundary and behavioural trajectory in which there are costs to changing activities is shown in Figure 4.1. The trajectory gives the optimal path in state space followed by a bird which can eat and drink in order to satiate its hunger and thirst. As the five arrows in the illustration show, such a bird begins by feeding until the trajectory encounters the dominance boundary, and then alternates between feeding and drinking until satiated. The triangles represent the set of possible arrows at each point in the trajectory as permitted by the experimental variables. The position of the dominance boundary is established empirically by interrupting ongoing behaviour and observing whether feeding or drinking occurs following the interruption. Note that the experimental use of interruptions to test this optimization model is different from that employed when examining time-sharing. Under time-sharing, one activity is assumed to be dominant to another, and interruptions help to detect this dominance. By contrast, under the optimization model, interruptions enable the location of a dominance boundary between two activities, each of which has a region of dominance within the entire state space. This contrast adds to the difficulty, pointed out earlier, of understanding a functional interpretation of time-sharing.

Ecology of motivation

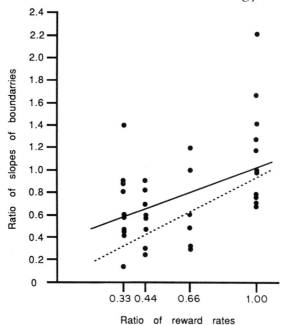

**Figure 4.2** The ratio of the slopes of the boundaries as a function of the ratio of the reward rates in doves. The broken line is based on the optimization theory while the solid line is a regression fitted to the data (from Sibly and McFarland, 1976).

The results of testing this optimization model with a quadratic cost function are given in Figure 4.2. In this figure the ratio of the slopes of boundaries for different water and food combinations are plotted against the ratio of the reward rates. The broken theoretical line is agreeably close to the solid regression line fitted to the data. The dominance boundary is stable under varying conditions of initial deprivation and the consequences of eating and drinking (Sibly and McCleery, 1976).

However, in his simulation model of this situation, discussed in section 2.5, Lester (1984b) found that the slope of the boundary is the ratio of the incentives only when expectations for food and water equal actual availabilities of those resources. Indeed, his model predicts the behaviour of doves during sessions in which the boundary is rotated by varying availabilities, better than does the one by Sibly (Figure 4.3). Thus, here is a model which proposes a proximate mechanism and produces output which is both empirically supported and theoretically reconciles optimal performance with the need to acquire information from the environment.

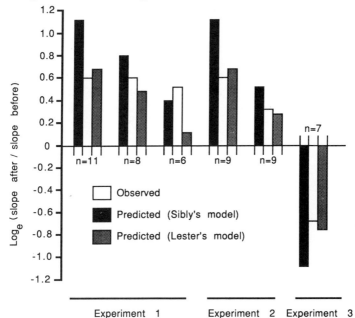

**Figure 4.3** Comparison of the fit of the models by Sibly (solid) and by Lester (hatched) with data (open) derived from three experiments in which dominance boundaries were rotated by varying availabilities (from Lester, 1984b).

## 4.2 MATCHING AND MAXIMIZING

Among the increasing number of topics on which the attention of behavioural biologists and psychologists has converged is that of how animals allocate their time and effort among alternatives (Lea, 1980; Staddon, 1980; Kamil and Roitblat, 1985). Parallels between, say, animals foraging under natural conditions, laboratory animals emitting operant responses for food reinforcements, and shoppers in a supermarket, are promoting interactions among ethologists, psychologists, and microeconomists. Such behaviour is of motivational interest because it can illustrate how underlying mechanisms monitor costs and benefits of time and energy. In this section we will examine the related topics of maximizing, matching, sampling and risk, and rules of thumb.

### Maximizing

From an adaptationist viewpoint, we might expect animals to maximize the rate at which they feed on different food types or in different food patches, or optimally choose times (give-up times) at which to leave these patches (reviews by Pyke, 1984; Stephens and Krebs, 1986; Kamil *et al.*, 1987) so that they can get on with other activities such as reproduction. This expectation is strengthened if feeding involves exposure to the risk of predation. Before proceeding, it should be noted that the utility of optimal theory has been challenged from various quarters. For instance, Pierce and Ollason (1987) raise eight objections which lead them to conclude that such theory is 'a complete waste of time'. However, most investigators (such as Stearns and Hempel (1987) in their reply to Pierce and Ollason), while acknowledging some difficulties, find the theory a useful framework for experimentation.

A variety of observations indicate that animals do manifest optimal behaviour in many situations. For instance, by using optimization calculations, Richard Cowie (1977) was able to predict quantitatively how much time great tits in an experimental feeding arrangement should spend in patches as a function of the interval required to travel between patches (Figure 4.4). The predictions when adjusted for

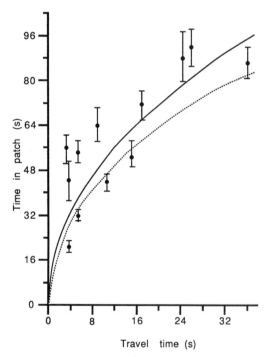

**Figure 4.4** Time spent in a food patch as a function of travel time between patches in a foraging experiment with great tits. The broken line represents a predicted optimal solution while the solid line represents a similar solution which includes energy expenditure (from Cowie, 1977).

energy expenditures were very good, and the results suggest that the birds can monitor both local and global rates of energy intake. Related work (Cowie and Krebs, 1979) indicated that the birds leave a patch according to a temporal rule of thumb under which they spend an amount of time there which is influenced by the relative richnesses of the available food patches.

In their ecological niches, animals are likely to be under a variety of constraints. For instance, matching provides a good description of the foraging behaviour of pied wagtails (*Motacilla alba yarrelli*) (Houston, 1986). However, the birds do not maximize their feeding rate, and show a bias toward being on their own territories. It may be that they are maximizing their chances of survival through this bias since sometimes only the territory offers feeding opportunities. Thus the bias stands as a constraint at the level of foraging, while possibly optimizing overall fitness.

## Matching

In contrast to maximizing, psychologists, particularly Richard Herrnstein and other operant researchers, have developed the concept of matching, attributed to the behaviour of animals responding on concurrent schedules of reinforcement under which the animal can make responses at more than one source of reward (e.g. DeCarlo, 1985) (Figure 4.5). For instance, a pigeon might have the choice of pecking a red key at which reinforcements are delivered every ten seconds on average after the previous reinforcement, or pecking a green key delivering reinforcements every twenty seconds on average. If the bird matches, then the proportion of pecks to one key ($R_1/(R_1 + R_2)$) is the same as the proportion of reinforcements at that key ($r/r_1 + r_2$) where $R$ = number of pecks and $r$ = number of reinforcements.

Although studied only in simplified form in the laboratory, matching and maximizing have been put forward as suggested mechanisms by which animals forage adaptively in the hurly burly of their natural environments. In many situations, the matching and maximizing hypotheses make identical predictions. This is easily shown for the concurrent variable-interval situation in which the animal distributes its total responses per time unit ($R$) between two alternatives, giving $pR$ ($= R_1$) to alternative 1 and $(1 - p)R$ ($= R_2$) to alternative 2 ($0 < p < 1$). With alternative 1 the mean interval between reinforcements ($t_1$) plus the mean interval when reinforcement is available but awaiting a response ($1/pR$) is the total mean time between reinforcements ($t_1 + (1/pR)$). Assuming instantaneously accessed reinforcements, the reciprocal ($(t_1 + (1/pR) - 1)$) gives the rate of reinforcement at alternative 1 ($r_1$). The total rate of reinforcement ($r_1 + r_2$)' is maximized when p = $r_1/(r_1 + r_2)$. In other

Ecology of motivation

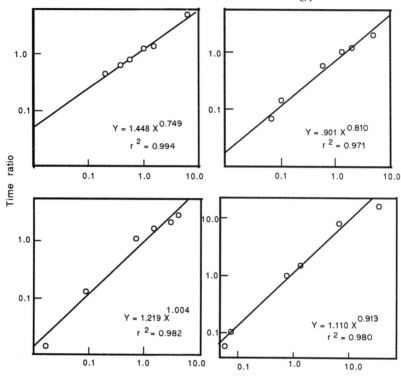

Reinforcement ratio

**Figure 4.5** Ratio of time at one of two concurrent operant schedules of reinforcement plotted against the reinforcement ratio in four pigeons (from DeCarlo, 1985).

words, when the derivative of the total rate of reinforcement with respect to $p$ is set equal to zero, i.e.

$$\frac{d}{dp}\,[r_1 + r_2] = \frac{d}{dp}\,[(t_1 + 1/pR) - 1 + (t_2 + 1/(1-p)R) - 1] = 0$$

then

$$p = r_1/(r_1 + r_2)$$

Since p also equals $R_1/(R_1 + R_2)$, maximization corresponds to matching. The general problem is to discover which of matching and maximizing, if either, is fundamental and which is a consequence.

## A crucial experiment

Mazur (1981) attempted an *experimentum crucis* involving pigeons pecking at coloured keys to distinguish between the hypotheses of

matching and maximizing. The concurrent variable-interval schedule of reinforcement included dark-key periods in which both red
and green keylights and the overhead chamber light were turned off.
In dark-key periods that provided reinforcement, grain was presented in an illuminated hopper while in non-reinforcement periods
the hopper was dark and empty. Such periods could be assigned to
either red or green keys by a timer. During the first phase of the
experiment, all dark-key periods included reinforcement, whereas in
the second phase only a random 10% of dark-key periods assigned to
the red key included reinforcement. Under this regime, matching
predicts a shift toward green-key responses whereas maximizing
does not. As Figure 4.6 shows, the results supported matching.
Mazur goes on to point out that the criticism that the pigeons did not
'understand' the reinforcement contingencies is invalid, since the
maximization hypothesis asserts that animals can monitor the rate of
reinforcement, and that this is all that matters.

Nevertheless, the experiment has been criticized as so artificial that
the mechanisms underlying performance become a constraint for
optimal performance (Krebs, 1983; Staddon and Hinson, 1983). Certainly it does not seem difficult to establish complex operant schedules which will produce unusual behaviour. Such schedules may
reveal operative mechanisms but not their normal function. In the
meantime, both operant psychologists and adaptationists carry on

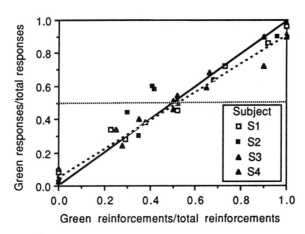

**Figure 4.6** Proportion of responses to the green key plotted against proportion of reinforcements at the green key in the second phase of Mazur's
(1981) experiment with pigeons. The broken line is a regression to the data
($y = 0.85x + 0.08$), the solid line is predicted by matching, and the dotted
line by maximizing.

happily. Among the psychologists, Green *et al.* (1983) interpreted results from experiments using concurrent ratio-interval schedules as supporting maximizing while other work has yielded results contrary to those expected under the hypothesis of maximizing (De-Carlo, 1985; Shettleworth, 1985). (Under a variable-ratio schedule, reinforcements are delivered after a variable number of responses.) Kamil and Yoerg (1982) have questioned whether the variable-interval schedules generally used in matching studies are valid models of the conditions under which most species feed in nature. More generally, the limited extent to which laboratory experiments successfully mimic natural foraging is a continuing source of concern. It is necessary wherever possible to corroborate such experiments with field observations.

## Sampling and risk

Both matching and maximizing models of performance require that the animal has knowledge of the food sources in the environment. The acquisition of this information is itself an intriguing problem. Operant psychologists have tended to gloss over such transients in performance and focus on steady-state output. Similarly, studies on foraging have generally reported the behaviour of informed animals, although 'sampling the environment' has been included to account for deviations from expected optimization. Nevertheless, some attention has been paid to sampling behaviour. Foraging at two patches with unknown food densities is similar to playing a two-armed bandit at a gambling casino, and indeed the behaviour of great tits (*Parus major*) in such a situation has been described with respect to the optimal solution to this problem (Krebs *et al.*, 1978). Just like a gambler, the birds in this experiment had to sample the resource patches before fixating on a patch to exploit. Krebs and his co-workers claimed a good agreement between their observations and predictions based on such a solution (Figure 4.7, solid line). In fact, as Plowright and Plowright (1987) have shown, the optimal solution is different (Figure 4.7, broken line) and the birds oversampled, at least when the difference in reward probabilities at the two patches was small. As the Plowrights point out, the subtleties of optimization models can be very great. More generally, as students are ruefully aware, behaviour is not optimal during learning, and optimal performance excludes learning: while learning, an animal adjusts its performance toward an optimal level, and once at an optimum the animal is no longer learning.

Food patches can provide the same average net benefits but differ in the variances of these benefits. Following economic terminology,

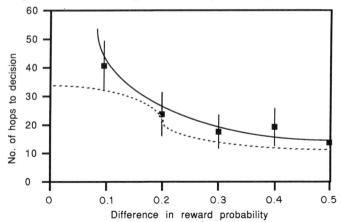

**Figure 4.7** Observed number of hops before deciding which of two food patches to exploit in great tits as a function of the difference in the reward probability at two patches. The lines are predicted from a two-armed bandit model for a session of 150 trails. The solid line is from Krebs *et al.* (1978) while the broken line is a correction from Plowright and Plowright (1987).

animals using a mechanism which ignores the variances are said to be risk-neutral, those in which it favours high-variance patches are risk-prone, and those in which it avoids such patches are risk-aversive. Different mechanisms can be more or less appropriate given the metabolic needs of the animal in its environment. Computer simulations indicate that when travel costs between patches are ignored, the RPS rule (described below) yields risk-neutral behaviour (Regelmann, 1986). When, more realistically, such costs are included, risk-aversive behaviour occurs, i.e. patches of constant reward are preferred over those with variable rewards. This is the case even though average net benefits are the same at the two patches. Empirical findings with bumble-bees, wasps and sticklebacks can be accounted for in terms of such an RPS rule. Results with shrews and passerine birds cannot be so interpreted. In these animals behaviour appears to depend on whether the necessary metabolic needs are being met. Thus the internal state of an animal influences the extent to which its foraging activity is risky.

## Rules of thumb

The problems of sampling the environment and of risk are routes to the motivational topic of the internal mechanisms which produce allocation of feeding efforts. It may well be that animals have been selected to use rules of thumb, or decision rules, which under many circumstances produce matching and maximization of reinforcement rate (Houston, 1983; Bookstaber and Langsam, 1985). (In economics, using inexpensive procedures which attain nearly optimal results is termed satisficing.) Rules of thumb may be more or less complex; may be inflexible or may vary with environmental conditions; and may involve learning. Several such rules of thumb have been put forth which can approximate optimal performance. Some of these rules, such as those embodied by the stochastic models of McNair (1980; 1981) do postulate learning. It is important to note that different species (e.g. dietary specialists versus generalists) may have different mechanisms, or at least mechanisms calibrated to different ecological situations. For instance, animals living in environments with unstable patches of food should change their estimates of patch profitability more quickly (i.e. use a shorter memory window) than those in environments with stable food patches.

Additionally, demonstrations that animals respond optimally to environmental variation do not necessarily provide insight into the operative causal system, which could be based on time, amount of food, or some combination of these expressed as a rate (Green, 1984). A discussion of this distinction is contained in McNamara and Houston's (1980) interesting interdisciplinary examination of the partial reinforcement effect using statistical decision theory. The partial reinforcement effect refers to the frequent finding that operant extinction is slower after partial reinforcement (i.e. reinforcement following only some responses) than after continuous reinforcement. This finding is counter-intuitive since, from an associationist viewpoint, it would be expected that a responding habit is stronger after continuous reinforcement and hence more resistant to extinction. A number of explanations have been put forward. For example, perhaps animals under partial reinforcement have more difficulty distinguishing acquisition and extinction. McNamara and Houston use Bayesian decision theory to show that it is optimal for animals to increase the persistence of their responding in the face of failure as the probability of a reinforcement decreases. They then go on to suggest a possible proximate mechanism which would produce this effect. The general point of importance is that animals can approximate optimal performance through the use of various rules of thumb.

What motivational machinery enables animals to learn about the changing availabilities of food resources in their environment? John Ollason (1980) has postulated a simple memory mechanism with which an animal can monitor its feeding rate and so decide when to remain in a food area and when to leave it. This model can be extended to predict spatial movements of foraging animals (Ollason, 1983). Similar models (Harley, 1981) have postulated a relative payoff sum (RPS) rule under which animals allocate time to patches in proportion to rewards gained there in the recent past. Expressed quantitatively, the probability of choosing the first of two patches at time $n + 1$ is $V_n(1)/(V_n(1) + V_n(2))$ where $V_n(1)$, the remembered value of patch 1 at time $n$, is $pV_{n-1}(1) + (1 - p)r_1 + F_n(1)$. The memory factor $p$ $(0 < p < 1)$ determines the decay of memory in time, $r_1$ is the pre-assigned value of the patch for the foraging animal, and $F_n(1)$ is the benefit from patch 1 actually obtained at time $n$. Thus patches of food are chosen in proportion to their values which are updated on the basis of previous foraging benefits. Such models have received support from experiments with foraging goldfish and stick-lebacks (Lester, 1984a; Milinksi, 1984). However, these models cannot account for feeding dynamics seen when patches are rapidly depleted (Kacelnik and Krebs, 1985).

Further on the question of proximate mechanisms which can track environmental changes in patchily distributed food, McNamara and Houston (1985) derived a rule, based on feeding rates and durations of visits in different patches, describing the manner in which an animal can learn about, and exploit, available resources. By includ-ing terms which discount information according to its recency, it is possible to pose a rule which is asymptotically optimal (i.e. produces optimal performance in the long run) and enables rapid shifts in behaviour in the face of environmental changes.

Also attending to the issue of rules of thumb, Hinson and Staddon (1983) have postulated that hill-climbing, i.e. picking the best option available from moment to moment, is the proximate means for pro-ducing both matching and overall maximizing. Similarly, based on a comparison of operant data with observations from studies in behav-ioural ecology, Fantino and Abarca (1985) have developed a delay-reduction hypothesis which asserts that reinforcers are effective to the extent that they predict a decrease in the time to food acquisition. This hypothesis is supported by a variety of findings dealing with such foraging aspects as selectivity among food types and temporal costs due to searching for and handling food. Recent experimental work (Ito and Fantino, 1986) has provided further support for the hypothesis. Based on earlier investigations by others, this study examined the performances of pigeons under an operant schedule intended to mimic natural foraging by including search, choice, and handling components. The results were consistent with both the delay-reduction hypothesis and the predictions of optimal for-aging theory.

Common to all of these suggested foraging mechanisms is an em-phasis on the monitoring of temporal aspects of food intake. Such monitoring implies ability to memorize information about the recent state of the environment. As an alternative to a memory window concept, Crawford (1983) has pointed to the operant phenomenon of local contrast as a possible proximate mechanism in foraging. Local contrast and induction, first described by Pavlov, refer to systematic variations in the response rate during stimulus A following stimulus B versus that following stimulus A itself. A change in response rate away from that under the previous stimulus is termed contrast, a change toward it is termed induction (see Figure 4.8). While there are parallels between expectations of behaviour under memory windows and local contrast, there are also differences. For instance, with a memory window, memory traces are assumed to decay faster than they accumulate, while local contrast predicts cumulative effects for easily discernible patches of food. Crawford has proposed several lines of research, such as manipulations of resource predictability, for investigating these possible mechanisms.

Overall, the matching or maximizing controversy has produced much interest in the causal mechanisms resulting in observed feeding performance. These rules of thumb have presumably been selected as

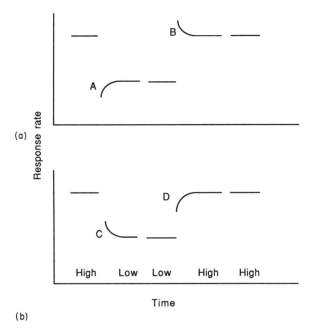

**Figure 4.8** Changes of response rate as a result of high and low reinforcement densities over time. (a) A and B represent negative and positive local contrast, respectively, and (b) C and D represent positive and negative induction (from Crawford, 1983).

ones which work well in most natural situations. Conversely, un-usual experimental situations may help to reveal the features of these rules. The results could well have consequences for our understand-ing of motivational processes generally by highlighting the impacts of general frameworks (mechanism, function), theoretical approaches (operant performance, behavioural ecology), and methodological procedures (such as simultaneous versus successive choice, or open versus closed economies). (In a closed economy, the study animals acquire all their food in the experimental situation through their own activity while, in an open economy, supplementary food outside this situation is available from the experimenter.) These diverse aspects of research into matching and maximizing indicate the manner in which understanding of motivation is advanced by the confluence of different approaches. Many issues remain to be resolved. For instance, a prediction of simple optimization models is that foraging animals should not exhibit partial preferences: each food type should be entirely included or excluded in the diet. In fact, animals often show partial preferences. Adaptationists have, not surprisingly, been eager and ingenious in their attempts to provide ultimate reasons for preferences (e.g. McNamara and Houston, 1987). It may be, however, that proximate accounts in terms of motivational dy-namics are more appropriate.

## 4.3 INDIVIDUAL DIFFERENCES

As Mayr (1982) has lucidly explained, the unfortunate legacy of typological thinking or philosophical essentialism (the argument that all entities are instances of various fixed, definitive essences) long led to a search for Platonic ideal forms in biology. Part of the Darwinian revolution was to produce a valid and necessary shift in attention to variation in characters, including behavioural ones. But it is only recently in ethology that such variation has been viewed as more than nuisance noise to be laundered away statistically. For instance, Machlis et al. (1985) show that, contrary to the common practice of pooling data across individuals, it is generally preferable to carry out analyses which assess variation within and between individuals separately.

Whenever behaviour is closely studied in any species, individual variation is encountered. We have already noticed this in yawning by male sticklebacks (Figure 2.1 p. 14), and Figure 4.9 provides a further example of such variation, in bluegill sunfish. Individual variation in a MAP can be assessed by focusing on the coefficient of variation, the unitless ratio of the standard deviation to the mean. As seen in instances such as features of chaffinch (*Fringilla coelebs*) song (Slater,

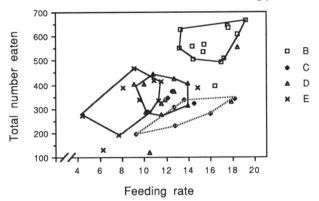

**Figure 4.9** Individual differences in total number eaten and feeding rate (based on the first 100 items eaten) in meals by four bluegill sunfish. The polygonal perimeters include at least 80% of the meals for each fish (from Lester, 1976).

1981), the coefficients of different behavioural characters are themselves quite variable. The intervals between successive notes produce coefficients of only a few %, whereas coefficients for the number of notes in a phrase are about 15%, and for the number of song types in a repertoire about 40%.

From a functional viewpoint (e.g. Clark and Ehlinger, 1987), it can be asked if such variation is adaptive or simply represents inevitable noise. Such a functional viewpoint can produce predictions for the patterning of life histories. From a motivational viewpoint, the central question is why different individuals, equal in such respects as size and sex, behave in consistently different ways. Obviously each individual has its own ontogeny and (except for clones) its own genome. As Hogan and Roper (1978) point out in a discussion of motivational systems, 'each system in each member of each species must be structurally unique, because the developmental history of no two individuals is identical'. The experimental task is to characterize this variation in its important dimensions.

In terms of feeding activities, individual specializations can arise from three sources: variable food supplies, phenotypic differences across individuals, and frequency-dependent payoffs (Partridge and Green, 1985). A patchy food supply can lead to changes in digestive physiology, improvements in the efficiency of feeding on particular foodstuffs with experience, and cultural inheritance due to established practices in the social group. In general, environmental unpredictability will prevent the optimization of one type of behaviour, and individual differences are likely to persist in a population.

Phenotypic differences have received little scrutiny, but promise to be a fruitful area of interaction between students of motivation and behavioural genetics in examining the hereditary basis of these differences. In zebra finches (*Taeniopygia guttata*), some individuals exhibit regular meal patterning while others exhibit short bouts with no regularity in length or onset (Slater, 1981). These differences may be a consequence of different social status.

Frequency-dependent payoffs can be interestingly approached using John Maynard Smith's (1982a) seminal concept of evolutionarily stable strategy (ESS). A strategy is a set of rules for a course of action in a game, that is, a competitive situation in which payoffs depend on the actions of the players. The strategy produces a particular payoff, measured in fitness, whose magnitude is a function of the frequencies with which other strategies are being played. In non-mathematical terms, an ESS is a strategy that, when commonly used in the population, has a larger payoff than any rival strategy. This framework may appear anthropomorphic but in fact is not, and is finding increasing application in ethology as well as evolutionary biology. In the case of feeding specializations, it is possible to show how different strategies enable individuals to maximize thier fitness. In reproductive behaviour, there are also alternative strategies, such as cuckoldry.

Individual variation in feeding can often be attributed to variation in experience with food types. This appears to be the case with Atlantic salmon (Marcotte and Browman, 1986). Such variation can have ecological consequences, affecting the population dynamics of both the species to which the varying individuals belong and that which constitutes the prey. Similarly, variation in predator avoidance, motor patterns, habitat use, and dominance hierarchies (e.g. Magurran, 1986) can be considered in terms of the three sources discussed above and the motivational mechanisms which play a role in producing this variation.

Like feeding behaviour, aggressive and defensive behaviour often manifests much individual variation. Such variation can be well described through the use of multivariate statistics. For example, comprehensive observations on the responses of sticklebacks to predaceous pike have been subjected to factor analysis which produces a number of independent dimensions from the overall data (Huntingford and Giles, 1987). In the case of the sticklebacks, eight dimensions reflected investigation of the pike, boldness towards it, three separate escape tactics, and three choices of microhabitat before the pike was presented (open water, beds of weeds, and cover on the bottom). Thus such a multivariate analysis facilitates the characterization of individual variation in defensive behaviour in prey species.

Aspects of individual variation in aggression between conspecifics will be discussed in section 4.5.

Beyond feeding and aggression, other functions of individual variation in social contexts include recognition as an individual (by fellow social members resulting in stable social relations), recognition of motivational state (Zahavi, 1980), or recognition of relatedness (important in kin selection). Individual recognition in flocks of black-capped chickadees (*Parus atricapillus*) is mediated through variation in such features of the song as frequency range and temporal duration (Mammen and Nowicki, 1981). There are also differences among flocks, and within a flock convergence of vocal performance occurs. Such convergence may facilitate the cohesion of the flock.

With respect to recognition of relations, an animal can enhance its overall, or inclusive, fitness if it can recognize them. This evolutionary process of kin selection, which underlies such societies as that of the bee hive, is examined in Colgan (1983) and Fletcher and Michener (1987). Each of the motivation, ontogeny, and function of individual differences will surely be the focus of increased research efforts in the future.

## 4.4  HUNGER AND FORAGING

Foraging is a topic of interest to behaviourists, community ecologists, and environmental physiologists. The behavioural aspect of foraging has several major components. For instance, learning plays an important role in many ecologically dominant vertebrates through such mechanisms as search images by which predators enhance their ability to detect prey over successive encounters or as adjustments in rates of searching for prey (Lawrence and Allen, 1983; Guilford and Dawkins, 1987; Kacelnik and Krebs 1985; and publications by Kamil). In the context of motivational analysis, it is interesting to examine the influence of hunger on foraging activities. Hunger, and its obverse, satiation, affect various aspects of feeding including searching, handling, and ratio measures of behaviour such as attacks per approach. As Dill (1983) has pointed out in an examination of these aspects in fish, such effects are adaptive in animals which must search out and capture prey because searching is increased and handling time is decreased when low food availability leads to hunger. In many species hunger depends on both gastric factors, reflected in stomach fullness, and central factors, representing metabolic state. Since deprivation for long periods is rare in nature, stomach fullness is usually an adequate measure of hunger. As satiation proceeds, searching decreases, ingestion per attack decreases, and handling time increases (Figure 4.10). These changes lead in hungrier animals

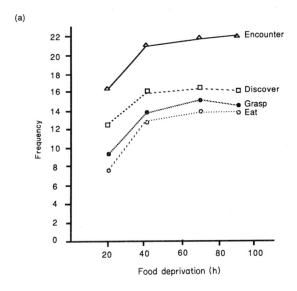

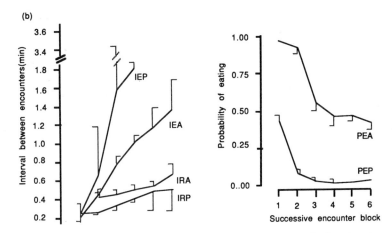

**Figure 4.10** Effects of hunger on foraging. (a) Food deprivation, by increasing hunger, increases the frequencies of encounters, discoveries, grasps, and eatings of prey in sticklebacks (from Beukema, 1968). (b) Satiation, by decreasing hunger, increases the interval between encounters and decreases the probability of eating prey consisting of adult or pupal flies in pumpkinseed sunfish. IEP: interval following eating a pupa; IRA: interval following refusing an adult; PEA: probability of eating an adult; etc. (from Colgan, 1973).

to higher attack rates and increased energy intake per unit handling time.

In an analysis similar to that by Dill, Elliott *et al.* (1977) examined prey capture by lions, and identified searching, stalking, attacking, and subduing phases of predation. They concluded that for a number of large mammalian predators hunger has its primary effect on the searching phase. Hunger varies with both the time since the last meal and the size of the meal. Indeed, this time is a linear function of the size of the meal. Hunger thus activates predators to enhance their searching for prey.

Hunger can affect aspects of foraging such as risk-aversion and diet breadth. In general, hungrier animals are less selective and more risk-aversive, showing preferences for food sources which have low variance. For example, pigeons responding in an operant situation show both decreased selectivity and a lower overall rate of reinforcement under greater deprivation (Snyderman, 1983). The author concludes 'that motivation has a great influence on prey selection, and probably on foraging behavior in general'.

The breadth of the diet of a feeding animal can be influenced by both external factors, such as the absolute and relative availabilities of food types, and internal factors contributing to motivational state. Richards (1983) has considered the role of satiation on optimal behaviour via diet breadth. It has been frequently observed that animals become more selective as hunger decreases. Richards' model predicts the reverse under some circumstances, especially if prey size is large relative to total intake. Standard optimal diet models calculate minimal foraging time given such features as energy values, handling times, and encounter rates of alternate prey. Richards adds the assumption that a predator can assess its hunger level, that is, the amount of food which it needs. The results for the mean and variance of foraging time indicate that a predator which expands its diet while feeding may be at an advantage over strict specialists or generalists under some conditions. These results merit support by experimental investigations.

In many cases, meal size and temporal patterning in feeding performance is dictated by the environment of the animal as much as by internal mechanisms. This realization leads to an ecological perspective on satiety (Collier, 1985). Two contrasting paradigms have been employed in the study of feeding. Under a session paradigm, a food-deprived animal eats a single meal, while under a free-feeding paradigm, a non-deprived animal eats *ad libitum*. The variables which control satiety, and the consequent behaviour exhibited by the animal, differ under each paradigm. Animals deal with a patchy environment through meal patterning and storage. The large storage capacities of lions and camels indicate responses to extreme patchiness. A variety of data indicate that when access to food becomes costly, meal size increases and meal frequency decreases. Cats maintained on laboratory operant schedules show these changes clearly (Figure 4.11). Across species there is diversity in the trade-off between meal size and frequency. Feeding rate over a meal does not decline in free-feeding animals, in contrast to those under a session paradigm. Like costs, benefits can also affect meal patterning. Feeding dynamics reflect the primary role of information acquired through external sources rather than internal feedback. Thus motivational analysis must include study of the mechanisms which animals use to acquire such information from a fluctuating and heterogeneous environment.

A variety of experiments by George Collier and co-workers illustrate this theme of the control of feeding by external cues. For instance, in a foraging experiment rats could gain access to diets differing in caloric density which were constantly available after a variable number of operant responses (Johnson et al., 1986). Body growth, daily intake of calories, and circadian patterning of feeding were similar across experimental diets. However, on the first day with a diet, animals on low-calorie diets ate more and longer meals, but nevertheless acquired fewer total calories compared with animals on high-calorie diets. Across diets, richer food was eaten faster. When changing between treatments, meal frequency changed quickly but meal size changed only over a few days on the new diet. Caloric intake was regulated over several meals rather than on a meal-to-meal basis.

In a further study (Johnson and Collier, 1987), rats chose between two foods differing in caloric value and availability as determined by the number of bar presses to gain a food pellet. The animals maintained roughly constant total daily caloric intake across experimental conditions. The diet selected was dependent on both of the foods available. Calorically dense foods were preferred, but this preference decreased as the density of the alternative food increased. Changing the availability of either food altered the diet chosen. For instance, a dense food was preferred only if its availability was not more than twice that of its alternative. The patterning and size of meals, and the rates of responding also varied as a function of the combination of foods offered.

Hunger can thus interact with external cues to influence foraging patterns in critical ways. This influence illustrates the necessity of examining links between motivational mechanisms and functional aspects of behaviour. It is also essential to investigate the rules of thumb, discussed in section 4.2, in order to understand how environmental information is evaluated in the context of the current

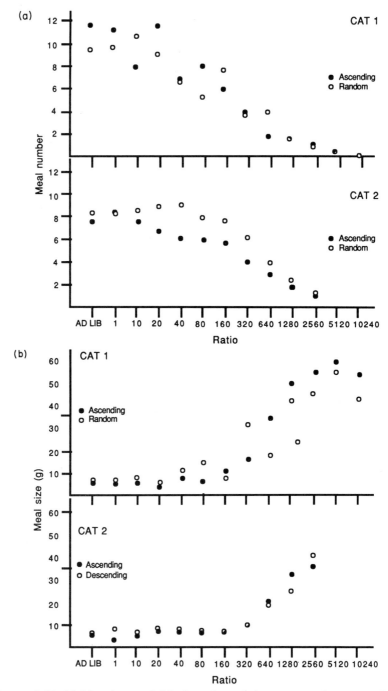

**Figure 4.11** (a) Number and (b) size of meals in cats as a function of the number of operant responses required for a food item. Ascending, descending, and random refer to the order of the operant schedules (from Collier, 1985).

metabolic state of the animal. Much remains to be learned about how environmental variables affect feeding. For instance, contrary to what might be expected, the effects of treatments in which the size of food items is varied do not parallel those in which caloric density is changed. Furthermore, correlations have often been sought be-between meal size and the length of pre-prandial or post-prandial interval as indicators of internal mechanisms regulating feeding. In fact, such correlations have only infrequently been found. With regard to optimal foraging theory, many studies on a number of species indicate that, contrary to the theory, animals do not maximize their feeding rate or show exclusive preferences by concentrating on the more profitable food types available. Overall, the relations of hunger and external cues in producing observed foraging behaviour merit more examination in species with diverse patterns of natural history.

## 5  SIGNALLING OF INTENTIONS

Students of motivation and communication (Sebeok, 1977) share common interests in the topic of signalling intentions. In the area of social communication (i.e. communication between conspecifics), there has been a long-standing interest in how actions, especially those resulting from conflict situations, evolve as signals between animals. In most species, agonistic encounters begin with antag-onists displaying tendencies to attack and to flee. Over the course of the encounter, the conflict is resolved and one opponent flees while the other remains. Displacement responses are especially common among territorial owners; perhaps because time is needed to assess the likely actions of intruders before the owners can decide on how to react (Halliday, 1980). Similar to aggression, courtship often in-volves opposing tendencies, with a gradual increase of sexual motiv-ation over aggression or fear. Baerends (1975) has evaluated the role of such conflict in the evolution of display in terms of identifying behavioural homologies and causal systems, relevant physiological data, and the variety of ethological observations that can be inter-preted in terms of conflict.

The hypothesis that conflicting motivational tendencies underlie much of display behaviour was prominent in the classical ethology of Lorenz and Tinbergen. The display vocalizations of nesting gulls and terns was cited as a source of support for this hypothesis. Re-cently Jan Veen (1987) has investigated these displays in great depth. For instance, in the little gull (*Larus minutus*), the alternation of notes of long and short duration in the long call reflect tendencies to attack and escape, respecitvely. Furthermore, the oblique posture of the body, the duration of the notes, the probability of an ensuing attack, and the relationship with the other bird all co-vary. Across the three relationships of offspring, mate, and non–related adult, both note duration and the probability of attack by the calling bird increase. This spectrum apparently represents an axis of increasing aggressive motivation. Veen concludes that such results corroborate the hypothesis that conflicting tendencies for attack and escape contribute to the production of display behaviour. The same display can result from a variety of motivational states, which can be distinguished by examining the elements (such as positions of limb components and feathers) of which the display consists. This finding explains why a particular display can be associated with a wide variety of attack and escape behaviour, and hence why it may be a poor indicator of ensuing activity. As discussed below under Intentions and ESSs, this lack of predictability is of theoretical importance in the analysis of communication.

The association of auditory social signals with underlying causal states in birds and mammals may reflect the operation of motivation-structural rules (Morton, 1977). Comparative data from numerous species suggest a structural convergence such that harsh (i.e. wide-band), low-frequency sounds indicate hostility while pure-tone, high-frequency sounds indicate fright or tendencies for appeasement or approach. Comparisons of signals within species also suggest this correlation between the structure of signals and their underlying motivation. Among many examples are the contact and aggressive calls of passerine birds. For instance, the Carolina chickadee (*Parus carolinensis*) produces sounds which are composite outcomes of hostile and friendly motivation. Based on such cases, it is possible to categorize calls, using features of band width and frequency modulation, as reflecting various combinations of aggression and appeasement. This analysis is similar to earlier ethological ones, such as the analysis of social postures in canids by Lorenz (1971). Thus the role of conflict in social communication is again revealed. The convergent basis of these motivation-structural rules may be that low, harsh sounds indicate body size, an important factor in aggression, while high-pitched, pure tones are associated with young which often require parental attention. This contrast hearkens back to Charles Darwin's (1872) Principle of Antithesis:

> Certain states of the mind lead . . . to certain habitual movements
> which were primarily, or may still be, of service; and we shall
> find that when a directly opposite state of mind is induced, there
> is a strong and involuntary tendency to the performance of
> movements of a directly opposite nature, though these have never
> been of any service.

Further analysis of the application of these rules to mammalian sounds (e.g. August and Anderson, 1987) are generally supportive. As expected, aggressive calls are generally low-pitched, but high-pitched sounds tend to be a motivational mixture of fearful and friendly states which require distinguishing, and fearful calls are very variable with respect to their frequencies. Overall, motivational considerations join usefully with the physics and physiology of signal production and ecological aspects of competition to provide understanding of the bases of social communication.

In this section we shall review the concepts of ritualization and emancipation, the honesty of signals, the use of evolutionarily stable strategies, and relevant data from cichlid fish.

## Ritualization and emancipation

Two central concepts for dealing with the evolution of displays are ritualization and emancipation. Ritualization is

> the adaptive formalization or canalization of emotionally
> motivated behaviour under the teleonomic pressure of natural
> selection so as: (a) to promote better and more unambiguous signal
> function, both intra- and inter-specifically; (b) to serve as more
> efficient stimulators or releasers of more efficient patterns of action
> in other individuals; (c) to reduce intraspecific damage; and (d) to
> serve as sexual or social bonding mechanisms (Huxley, 1966).

Ritualization leads to stylized, exaggerated performance of actions as seen in human rituals. This process was first described by Julian Huxley in his classical report on courtship in the great crested grebe, and has played a major role in understanding the evolution of animal signals. Emancipation refers to the associated process through which the causal factors of such signals can change phylogenetically as selection acts on them (Tinbergen, 1952). The display vocalizations of gulls and terns, studied by Veen (1987) and discussed above, reflect emancipation. Among vertebrates many agonistic and courting signals are regarded as being derived from, for instance, displacement acts of grooming, and to have evolved to become elements of aggressive and sexual systems. The evolution of display behaviour is often closely associated with that of the anatomical features (such as crests or flaps) which are employed in the display.

Ritualization is the process by which signals evolve from non-signals. As defined by Huxley, ritualization emphasizes co-operation between interacting individuals through the use of unambiguous signals for the speedy resolution of conflicts or courtships for the benefit of all involved. In the same vein, Amotz Zahavi (1980) has elaborated Morris's concept of typical intensity (discussed in Chapter 2) and has suggested that competition among conspecifics for the clear recognition of social signals drives the process of ritualization. By contrast, more recent considerations of communication (e.g. Krebs and Dawkins, 1984) stress that signalling can be expected to evolve as a benefit to the signaller and that this evolution may involve manipulation and deception of conspecific recipients, be they mates or rivals, just as of non-conspecifics, as in mimicry and camouflage. A signalling animal manipulates a rival to flee or a sexual partner to mate, while recipients 'mind-read', i.e. predict what that animal will do next based on those signals. It may, or may not, be advantageous for a signalling animal to have its mind read. In the latter case, manipulation and mind-reading co-evolve in an arms race, with signals as the evolutionary products of this race. The detection of whether any particular social communication is co-operative or exploitative can perhaps be accomplished by studying the amplitude and conspicuousness of signals involved. Since there are costs associated with signalling, co-operative signalling may evolve to the use of low amplitude, and hence inexpensive, signals, whereas exploitation involves hard selling with large advertisements. Additionally, it is necessary to include consideration of ecological constraints such as social spacing, as well as other advantages, such as discretion while co-operating.

## Honesty of signals

The issue of signalling intentions is a component of the larger topic of honest and deceptive signals in animal communication generally. Individuals may withhold or falsify information both interspecifically and intraspecifically. Between species such communication involves camouflage, mimicry, and the feigning of death or injury in the presence of attackers. (In an oft-cited anecdote (e.g. Campbell, 1982) Louis Leakey claims that African hares, when pursued, fold back their ears immediately prior to dodging to one side, and that this action can be detected by a predator who can profit by it. It is difficult to see any advantage to this behaviour.) Within species, it involves displays associated with aggression and courtship (including sexual mimicry and deception with respect to resources which are evaluated by potential mates), as well as signals such as alarm calls and feeding calls associated with the sharing of food and dependent on the quality of food and on which other animals are within hearing range.

The honesty of signals has been the subject of numerous studies, as for example in status signalling (Rohwer and Ewald, 1981) and dialect use in birds (Baker and Cunningham, 1985). In the latter case, the causes of vocal dialects have been hotly debated. It has been suggested, for instance, that they are the result of geographic isolation or that they contribute to the local adaptation of populations. By contrast, Rothstein and Fleischer (1987) argue that dialects serve as honest signals of the quality of males as mates. Among western populations of the North American brown-headed cowbirds (*Molothrus ater*), males produce a flight whistle immediately prior to copulation which is dialectally variable and which probably influences mate choice by females. Females of this brood-parasitic species prefer to mate with males of high status, which accrues with age. A reflection of age is the production of the whistle of the local dialect. The signalling is honest because males who are young or emigrate, and therefore have not been resident long enough to gain social dominance, cannot produce the local dialect. These dialects are thus maintained by a process of honest convergence to the local form of the flight whistle which attracts mates.

Whether signalling is honest or not, animals may consistently produce a given signal in a particular context or, alternatively, may exhibit behavioural flexibility according to the social situation. It is this latter flexibility, and particularly the signalling of intentions, for which motivational analysis is of paramount importance. Intentionality has long received much attention from philosophers of mind (e.g. Diamond and Teichman, 1979; Dennett, 1983) and is central in legal proceedings to establish guilt. Some sense of awareness seems to be intrinsic to intentionality. There is no generally accepted operational definition of intention at any level of analysis and hence the scientific use of this term is unclear. Nevertheless, students of animal behaviour have dealt with the issue because the swagger and bravado seen in courtship and combat give pause to wonder about the honesty of the signalling. In his discussion of the risks of lying, W.J. Smith (1977) concluded that 'ethologists have not yet caught non-human animals in intentional acts of lying with display behavior'. Recent findings do seem to indicate mendacity. Koko, a gorilla tutored in sign language, is reported to have lied (Patterson, 1978). In the intensively studied three-spined stickleback, courtship may involve considerable deception (Rohwer, 1978). In this species, nesting males aerate eggs by fanning. Fanning may be performed deceptively by males during courtship to convince females that eggs are already present and that the male is therefore a desirable mate. (Males generally fast during nesting, but may eat eggs. Thus the presence of eggs in a nest may indicate to a female that her eggs are at low risk of being eaten.) Beyond experimental data, purveyors of anecdotes are always prepared to rise on their hind legs with many alleged instances of lying available.

## Intentions and ESSs

By what means could deceitful communication evolve? For instance, should animals make 'big lies' (Wallace, 1973)? Using the concept of evolutionarily stable strategy, it can be asked whether it is an ESS for an animal to signal honestly about its size, strength, or intentions. Depending on assumptions about such matters as the ability for individual recognition (Rhijn and Vodegel, 1980), it is possible to set up many different models (Rhijn, 1980). In ESS models of aggressive interactions it has proven useful to distinguish the resource holding potential (RHP) of an individual, which is associated with its physical strength and competence in battle, and its knowledge about the value ($V$) of a resource to both itself and to competitors (Parker, 1984). Differences in $V$ can affect the motivation of an animal to defend a resource, and so produce asymmetric contexts. This concept of the value of a resource thus provides a link between proximate and ultimate aspects of aggression.

Peter Caryl (1979) has considered animal communication from an ESS viewpoint and contrasted this approach with that of classical ethology in which a mutually beneficial sharing of information was envisioned. He concluded that many available data are usefully interpreted with models based on ESSs and that, contrary to the tenets of ethology, the transfer of information in aggressive encounters is poor. In response, Robert Hinde (1981) has argued that ESS proponents have caricatured classical ethology and set up straw men, particularly on the issues of selection pressures and the information content of displays. On the latter point, well-established observations that displays have typical intensities and can be followed by a number of responses by signalled animals show that ethologists have long been aware that information content of displays is low. Hinde points out that these observations are compatible with Maynard Smith's suggestion that typical intensities have evolved to conceal what the animal will do next. He indicates where further data on the motivational basis of animal signalling are needed, and suggests that an ESS approach can augment traditional ethological investigation. However, Caryl (1982a) has reiterated his initial criticisms of the ethological emphasis on unambiguous exchanges of information and his conclusions on the usefulness of ESS models for examining data. These viewpoints serve to highlight the different meanings of the term information, as in its comparative or theoretical senses, and the consequent assortment of orientations taken by researchers.

A prediction arising from ESS models is that animals, while exchanging information on size and strength, should not generally give information about their intentions. Moynihan (1982) has challenged this prediction by arguing that intentions are signalled by both ritualized and unritualized displays, and that these messages are generally honest. He has gone on to suggest that perhaps these signals are sceptically evaluated by recipients and thus are the only truthful ones in animal communication. In their replies, Maynard Smith (1982b) and Caryl (1982b) emphasize the importance of distinguishing between the following:

1. ritualized displays from unritualized movements;
2. what an animal signals from what a recipient can detect;
3. RHP, which is determined by such factors as size which are not easily increased, from *V*, which can be.

Maynard Smith points out that the ESS analyses have been based on ritualized displays, and that these show that honest signalling is ESS for RHP since lying is not possible, whereas this is not the case with intentions. Caryl has gone on to show that a variety of observations support the idea of sceptical recipients as incorporated in an ESS model. Available data indicate that the behaviour of opponents is no better predicted by unritualized movements than it is by ritualized display, and that it is necessary to examine the nature of the interactions.

## Relevant data from cichlid fish

A number of studies on aggressive behaviour in fish have provided data which bear on the issue of concealing intentions. One such recent experiment (Turner and Huntingford, 1986) examined staged contests between males of the Mozambique mouthbrooder (*Oreochromis mossambicus*). The contests consist of circling, lateral displays, and various types of approach and contact, and they end when the losing fish dashes away or submits to his opponent. The larger fish almost always wins, and information about relative size is apparently acquired during the interactions. Size discrepancy is not associated with the length of a contest as measured in terms of time or acts. There is only a weak negative association between discrepancy in body size and intensity of contest as measured by acts per minute. Pertinent to a discussion of intentions is the finding that the behaviour of winners and losers is different in several respects throughout the contest. Winners circle around and tailbeat more than losers, and show a higher rate of rams (in which the opponent is contacted) to charges (in which he is not) (Figure 4.12). These behavioural differences are found distributed across the duration of the contest but increase as the contest progresses. The discovery of these differences

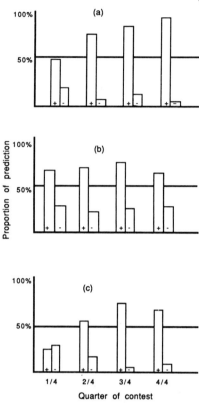

**Figure 4.12** Prediction of outcome of contest over successive quarters based on (a) circling, (b) tailbeating, and (c) ratio of rams to charges in cichlid fish. Predictions are based on cumulated occurrence of the activity by each contestant, with ties omitted. Positive and negative refer to the correctness or incorrectness, respectively, of the prediction as to the outcome of the contest based on the activity up to that quarter (from Turner and Huntingford, 1986).

is an advance over earlier research which failed to demonstrate them for a variety of reasons.

During the contests of these fish, as in many animals, information about size and strength and about intentions is often unavoidably jointly transmitted in the performance of individual acts. When various acts can be employed during a contest, the choices made by the animal can reflect its intentions more clearly. As the interaction proceeds, each contestant can modify its signals regarding size and intentions, and concurrently receive information on those features from its opponent. Functionally, benefits of being relatively large or

energetic and of bluffing must be offset by costs of being small or
fatigued and of having a bluff called. Additionally, it is important to
consider the time scale which is operative. For instance, intention
movements are short-term predictions of behaviour likely to occur
in the immediate future. At the other extreme, activities which are
useful as predictors of the eventual outcome of a contest are not
conventionally considered signals of an animal's intentions to lose or
win in the end. Activities may contain information on intentions
because such information is not detected by its opponent and therefore there has been no selection against broadcasting this information. Further, the cost of providing this information may be weighed
against some greater gain in information acquired from the response
of an opponent. Finally, selection pressures from, say, predators
may select for economical and honest signalling of intentions as well
as of size and strength. Thus the requirement to conceal intentions is
only one of numerous requirements which must be met through
compromise in animal communication systems.

A second recent experiment compared the aggressive behaviour of
Midas cichlids (*Cichlasoma citrinellum*) toward conspecific opponents
and surrogate stimuli (Barlow *et al.*, 1986). These cichlids showed a
high degree of individual variation in aggressive scores when tested
against such stimuli. After this testing, paired fish were held in visual
isolation in either half of an experimental aquarium for either 1–2
hours or 24 hours before contact was allowed. Contact involved
charging, circling, and biting. After the longer isolation, the only
prediction of victory was the size of the fish. The difference in size
of the fish did not correlate with the duration of the contests. After
the shorter isolation, the aggressive scores, but not body size, predicted winners of contests. Compared with contests after the longer
isolation, these contests were briefer and developed more slowly.
Surprisingly, the more similar the aggressive scores of the pair of
fish, the briefer the display phase of the contest. The authors suggest
that after the longer isolation the fish were behaving more territo-
rially and that the contests reflect prowess based on size while after
the shorter isolation the contests reflected 'daring' based on aggressiveness. Further research can illuminate the extent of the motivational flexibility of this daring.

Signalling intentions is clearly a topic to attract the attention of
cognitive ethologists such as Ristau (1983) whose research on the
broken-wing displays of nesting plovers was discussed in section
2.7. Indeed, this topic is attracting the attention of behaviourists,
cognitivists, evolutionary biologists and philosophers in a most exciting fashion.

# Overview

Like Churchill's Russia, the topic of motivation has struck many as a riddle wrapped in a mystery inside an enigma. Motivational analysis is not easy, and there is truth behind the humour of Dubos' (1971) conclusion that 'under precisely controlled conditions an animal does as he damn pleases'. It is worthwhile emphasizing the chief conclusions of this survey of motivational research by focusing on structure and function; causal versus functional analysis; terminological clarity at each level of examination; key principles; and quantitative modelling. The causal structure and dynamics of animal behaviour are clear in their chief features. Structurally animals are equipped with a set of causal systems which deal with the major behavioural tasks which they encounter. The systems consist of a hierachy of acts, actions, and activities, and also include internal feedback loops as well as cross-connections with each other. Jointly, they determine the choices among the available options. Generally, transitions between successive activities are smooth, but conflict is occasionally extended. These dynamics result from external stimuli and internal drives. The stimuli direct orientations, release consummatory responses, act as incentives to appetitive performance, and prime various systems. The drives can either have effects specific to one system, or are more general in arousing and directing responses. The question of whether drives are specific or general disappears as do similar questions about whether the control for behaviour is central or peripheral, whether acts are learned or innate, or whether ethology or psychology is a superior discipline.

Amid the communal cacophony of physiologists, ethologists, and ecologists, it is important to realize the separateness of causation and function: causal analysis does not indicate adaptiveness, and functional analysis does not reveal causal mechanism. In particular, hypotheses about causation and those about function are not mutually exclusive, as exemplified by the work on matching and maximizing in feeding behaviour.

Without venturing too deeply into ontological and epistemological morasses, it is clear that motivational terminology remains a cluttered vocabulary. There are various ways of describing specific acts (by topography, by causation, and by function), of summarizing sets of actions related in some respect (by form, time, cause, or role), and of expressing hypothetical entities within some theoretical framework. The cleaning of these Augean stables is an essential task, with an aim to establishing a minimal yet comprehensive language in which to deal with these problems. Terminological clarity will be assisted by attention to the operative level of analysis: physiological, ethological, or functional. The foregoing material outlines the concepts and tools appropriate to each level, and the contribution which each makes to an understanding of motivation. For instance, physiological studies on the effects of electrical stimulation of the rat brain reveal stimulus-bound feeding or drinking, depending on whether food or water is available. The ethological parallel is the recurrence of displacement responses which can be influenced by similar external stimuli. These two sets of observations jointly reveal the role of cues in directing aroused animals. Similarly, priming is an important physiological process and has also long been a central ethological concept, as evidenced by its inclusion in Tinbergen's classic definition of instinct.

Linking these levels are certain key principles. A major one of these is hierarchical organization, a principle which is dominant throughout biology (O'Neill et al., 1986). Physiological findings indicate the proximate basis of this organization, including the importance of command centres, feedback loops, and the mutual influence of components. Ethological data illustrate the interactions within and among these behavioural systems in establishing priorities for activities. Functional data provide insights into the evolutionary pressures which shape efficient hierarchical systems. Hierarchy also provides a means of moving across levels to integrate, for example, the central excitatory states studied by behavioural physiologists with the state spaces of ethologists.

If these conclusions are sound, then the role of quantitative modelling will inevitably become larger in motivational research. Current attempts as witnessed in optimization and catastrophe theory may be clumsy or inadequate, but they are surely steps in the right direction. Expertise in these topics will certainly be a prerequisite for contributors. Eventually, if Pythagoras and his followers are correct in maintaining that all things are number, then ethometric models, the $n$th generation successors of the quantitative work reviewed above, should link these levels of analysis, dispel terminological disputes, and provide a comprehensive basis for understanding motivation. In these models the terms will be operationally linked to data while the structure of the equations will yield functional features and embody the meaning of general principles such as regulation, competition, inhibition, hierarchy, feedback, and priming, as well as specific cognitive concepts. As Leibniz envisaged with his universal language, the beatific researchers of that era will simply calculate.

# References

Achinstein, P. and Barker, S.F. (1969) *The Legacy of Logical Positivism*. Johns Hopkins University Press, Baltimore

Ackroff, K., Schwartz, D. and Collier, G. (1986) Macronutrient selection by foraging rats. *Physiol. Beh.*, **38**, 78–80

Armstrong, E.A. (1947) *Bird Display and Behaviour, 2nd ed.* Drummond, London

Aschoff, J. (ed.) (1981) *Handbook of Behavioral Neurobiology. Vol. 4, Biological Rhythms*. Plenum Press, New York

Ashby, W.R. (1952) *Design for a Brain*. Chapman and Hall, London

Atkinson, J.W. (1964) *An Introduction to Motivation*. Van Nostrand Reinhold, New York

Atkinson, J.W. and Birch, D. (1970) *The Dynamics of Action*. Wiley, New York

August, P.V. and Anderson, J.G.T. (1987) Mammal sounds and motivation structural rules: A test of the hypothesis. *J. Mamm.*, **68**, 1–9

Baerends, G.P. (1975) An evaluation of the conflict hypothesis as an explanatory principle for the evaluation of displays. In *Function and Evolution of Behaviour* (eds G.P. Baerends *et al.*), Oxford University Press, Oxford, pp. 187–227

Baerends, G.P. (1984) The organization of the pre-spawning behavior in the cichlid fish *Aequidens portalegrensis* (Hensel). *Neth. J. Zool.*, **34**, 233–366

Baerends, G.P., Brouwer, R. and Waterbolk, H.T. (1955) Ethological studies on *Lebistes reticulatus* (Peters). 1. An analysis of the male courtship pattern. *Behaviour*, **8**, 249–334

Baerends, G.P. and Kruijt, J.P. (1973) Stimulus selection. In *Constraints on Learning* (eds R.A. Hinde and J. Stevenson-Hinde), Academic Press, London, pp. 23–50

Baerends, G.P. and Drent, R. (1982) The herring gull and its eggs. *Behaviour*, **82**, 1–416

Baker, M.C. and Cunningham, M.A. (1985) The biology of bird-song dialects. *Beh. Brain Sci.*, **8**, 85–133

Bambridge, R. and Gijsbers, K. (1977) The role of tonic neural activity in motivational processes. *Exp. Neurol.*, **56**, 370–85

Barlow, G.W. (1977) Modal action patterns. In *How Animals Communicate* (ed T.A. Sebeok), Indiana University Press, Bloomington, pp. 98–134

Barlow, G.W., Rogers, W. and Fraley, N. (1986) Do Midas cichlids win through prowess or daring? It depends. *Beh. Ecol. Sociobiol.*, **19**, 1–8

Barnes, W.J.P. and Gladden, M.H. (1985) *Feedback and Motor Control in Invertebrates and Vertebrates*, Croom Helm, London

Bateson, P.P.G. and Klopfer P.H. (eds) (1982) *Perspectives in Ethology. Vol. 5 Ontogeny*, Plenum Press, New York

Bell, W.J. (1989) *Searching Behaviour*, Chapman and Hall, London

Bernard, C. (1927) *An Introduction to the Study of Experimental Medicine*, Macmillan, London

Bertalanffy, L.V. (1968) *General System Theory. Rev. ed.*, Braziller, New York

Berthold, A.A. (1849) Transplantation der Hoden. *Arch. Anat. Physiol. Wissensch. Med.*, **16**, 42–6

Berthoud, H.R. and Powley, T.L. (1985) Altered plasma insulin and glucose after obesity-producing bipiperidyl brain lesions. *Am. J. Physiol.*, **248R**, 46–53

Beukema, J. (1968) Predation by the three-spined stickleback (*Gasterosteus aculeatus* L.): The influence of hunger and experience. *Behaviour*, **31**, 1–126

Bindra, D. (1959) *Motivation*, Ronald, New York

Bindra, D. (1978) How adaptive behavior is produced: A perceptual–motivational alternative to response-reinforcement. *Beh. Brain Sci.*, **1**, 41–91

Boden, M. (1981) *Minds and Mechanisms*, Cornell University Press, Ithaca

Bolles, R.C. (1975) *Theory of Motivation. 2nd ed.*, Harper & Row, New York

Bonsall, R.W., Zumpe, D. and Michael, R.P. (1978) Menstrual cycle influences on operant behavior of female rhesus monkeys. *J. Comp. Physiol. Psychol.*, **92**, 846–55

Bookstaber, R. and Langsam, J. (1985) On the optimality of coarse behavior rules. *J. Theor. Biol.*, **116**, 161–93

Bowdan, E. and Dethier, V.G. (1986) Coordination of a dual inhibitory system regulating feeding in the blowfly. *J. Comp. Physiol.*, **158A**, 713–22

Brenner, S. (1974) The genetics of *Caenorhabditis elegans*. *Genetics*, **77**, 71–94

Bridgman, P.W. (1950) *Reflections of a Physicist*, Philosophical Library, New York

Broom, D.M. (1968) Behaviour of undisturbed 1- to 10-day-old chicks in different rearing conditions. *Devel. Psychobiol.*, **1**, 287–95

Broom, D.M. (1979) Methods of detecting and analysing activity rhythms. *Biol. Beh.*, **4**, 3–18

Broom, D.M. (1980) Activity rhythms and position preferences of domestic chicks which can see a moving object. *Anim. Behav.*, **28**, 201–11

Brown, J.A. and Colgan, P.W. (1985) Interspecific differences in the ontogeny of feeding behaviour in four species of centrarchid fish. *Zeits. Tierpsychol.*, **70**, 70–80

Brown, J.S. (1961) *The Motivation of Behavior*, McGraw-Hill, New York

Brown, R.E. and McFarland D.J. (1979) Interaction of hunger and sexual motivation in the male rat: A time-sharing approach. *Anim. Behav.*, **27**, 887–96

Bunge, M. (1977) Emergence and the mind. *Neuroscience*, **2**, 501–9

Cabanac, M. (1979) Sensory pleasure. *Quart. Rev. Biol.*, **54**, 1–29

Calow, P. (1976) *Biological Machines A Cybernetic Approach to Life*, Edward Arnold, London

Campbell, B.G. (ed.) (1982) *Humankind Emerging*, 3rd ed., Little Brown, Boston

Campfield, L.A., Smith, F.J. and Fung, K.F. (1982) Ventromedial hypothalamus hyperphagia and obesity: Role of autonomic neural control of insulin secretion. In *The Neural Basis of Feeding and Reward* (eds B.G. Hoebel and D. Novin), Haer Institute, Brunswick, pp. 203–20

Cannon, W.B. (1919) *Bodily Changes in Pain, Hunger, Fear and Rage*, Appleton-Century, New York

Caryl, P.G. (1979) Communication by agonistic displays: what can games theory contribute to ethology? *Behaviour*, **68**, 136–69

Caryl, P.G. (1982a) Animal signals: A reply to Hinde. *Anim. Beh.*, **30**, 240–4

Caryl, P.G. (1982b) Telling the truth about intentions. *J. Theor. Biol.*, **97**, 679–89

Cassidy, J. (1979) Half a century on the concepts of innateness and instinct: Survey, synthesis, and philosophical implications. *Zeits. Tierpsychol.*, **50**, 364–86

Churchland, P.S. (1986) *Neurophilosophy*, Massachusetts Institute of Technology, Cambridge, MA

Clark, A.B. and Ehlinger, T.J. (1987) Pattern and adaptation in individual behavioural differences. In *Perspectives in Ethology, Vol. 7* (eds P.P.G. Bateson and P.H. Klopfer), Plenum Press, New York, pp. 1–47.

Cofer, C.N. and Appley, M.H. (1964) *Motivation: Theory and Research*, Wiley, New York

Cohen, J. (1975) Cultural homology. *Science*, **187**, 907–8

Cohen, S. and McFarland, D. (1979) Time-sharing as a mechanism for the control of behaviour sequences during the courtship of the three-spined stickleback (*Gasterosteus aculeatus*). *Anim. Beh.*, **27**, 270–83

Colgan, P.W. (1973) Motivational analysis of fish feeding. *Behaviour*, **45**, 38–66

Colgan, P.W. (ed.) (1978) *Quantitative Ethology*, Wiley, New York

Colgan, P.W. (1983) *Comparative Social Recognition*, Wiley, New York

Colgan, P.W. (1986) Motivational basis of fish behaviour. In *The Behaviour of Teleost Fishes* (ed. T.J. Pitcher), Croom Helm, London, pp. 23–46

Colgan, P.W. and Gross M.R. (1977) Dynamics of aggression in male pumpkinseed sunfish (*Lepomis gibbosus*) over the reproductive phase. *Zeits. Tierpsychol.*, **43**, 139–51

Colgan, P.W. and Slater, P.J.B. (1980) Clustering acts from transition matrices. *Anim. Beh.*, **28**, 965–6

Colgan, P.W., Nowell, W.A. and Stokes, N.W. (1981) Nest defense by male pumpkinseed sunfish (*Lepomis gibbosus*): Stimulus features and an application of catastrophe theory. *Anim. Beh.*, **29**, 433–42

Collier, G.H. (1980) An ecological analysis of motivation. In *Analysis of Motivational Processes* (eds F.M. Toates and T.M. Halliday), Academic Press, London, pp. 125–51

Collier, G.H. (1982) Determinants of choice. *Nebraska Symp. Motivation 1981*, 69–127

Collier, G.H. (1985) Satiety: An ecological perspective. *Brain Res. Bull.*, **14**, 693–700

Collier, G.H. and Rovee-Collier, C.K. (1983) An ecological perspective of reinforcement and motivation. In *Handbook of Behavioral Neurobiology Vol. 6 Motivation* (eds E. Satinoff and P. Teitelbaum), Plenum Press, New York, pp. 427–41

Cowie, R.J. (1977) Optimal foraging in great tits (*Parus major*). *Nature*, **268**, 137–9

Cowie, R.J. and Krebs, J.R. (1979) Optimal foraging in patchy environments. In *Population Dynamics* (eds R.M. Anderson *et al.*), Blackwell, Oxford, pp. 183–205

Cox, J.E. and Smith, G.P. (1986) Sham feeding in rats after ventromedial hypothalamic lesions and vagotomy. *Beh. Neurosci.*, **100**, 57–63

Crawford, L.I. (1983) Local contrast and memory windows as proximate foraging mechanisms. *Zeits. Tierpsychol.*, **63**, 283–93

Crespi, L.P. (1942) Quantitative variation of incentive and performance in the white rat. *Am. J. Psychol.*, **55**, 467–517

Crews, D. (1979) Neuroendocrinology of lizard reproduction. *Biol. Reprod.*, **20**, 51–73

Crews, D. (1984) Gamete production, sex hormone secretion, and mating behavior uncoupled. *Horm. Beh.*, **18**, 22–8

Croll, R.P. and Davis, W.J. (1981) Motor program switching in *Pleurobranchaea*. *J. Comp. Physiol.*, **145**, 277–87

Darwin, C.R. (1872) *The Expression of the Emotions in Man and the Animals*, Murray, London

Davis, W.J. (1984) Motivation and learning: Neurophysiological mechanisms in a 'model' system. *Learn. Motiv.*, **15**, 377–93

Davis, W.J. (1985) Neural mechanisms of behavioural plasticity in an invertebrate model system. In *Model Neural Networks and Behavior* (ed. A.I. Selverston), Plenum Press, New York, pp. 263–82

Dawkins, M.S. (1980) *Animal Suffering*, Chapman and Hall, London

Dawkins, R. (1976) Hierarchical organisation: A candidate principle for ethology. In *Growing Points in Ethology* (ed. P.P.G. Bateson and R.A. Hinde), Cambridge University Press, Cambridge, pp. 7–54

DeCarlo, L.T. (1985) Matching and maximizing with variable-time schedules. *J. Exp. Anal. Beh.*, **43**, 75–81

Dennett, D.C. (1983) Intentional systems in cognitive ethology: The 'Panglossian paradigm' defended. *Beh. Brain Sci.*, **6**, 343–90

Dethier, V.G. (1976) *The Hungry Fly*, Harvard University Press, Cambridge, MA

Dethier, V.G. (1982) The contribution of insects to the study of motivation. In *Changing Concepts of the Nervous System* (eds A.R. Morrison and P.L. Strick), Academic Press, New York, pp. 445–55

Diamond, C. and Teichman, J. (eds) (1979) *Intention and Intentionality*, Harvester Press, Brighton

Dickinson, A. (1980) *Contemporary Animal Learning Theory*, Cambridge University Press, Cambridge

Dill, L.M. (1983) Adaptive flexibility in the foraging behavior of fishes. *Can. J. Fish. Aq. Sci.*, **40**, 398–408

Douglas, J.M. and Tweed, R.L. (1979) Analysing the patterning of sequence of discrete behavioural events. *Anim. Beh.*, **27**, 1236–52

Drummond, H. (1981) The nature and description of behaviour patterns. In *Perspectives in Ethology* (eds P.P.G. Bateson and P.H. Klopfer), Plenum Press, New York, pp. 1–33

Dubos, R. (1971) In defense of biological freedom. In *The Biopsychology of Development* (eds E. Tobach *et al.*), Academic Press, New York, pp. 553–560

Duggan, J.P. and Booth, D.A. (1986) Obesity, overeating, and rapid gastric emptying in rats with ventromedial hypothalamic lesions. *Science*, **231**, 609–11

Elliott, J.P., Cowan, I.McT. and Holling, C.S. (1977) Prey capture by the African lion. *Can J. Zool.*, **55**, 1811–28

Epstein, A.N. (1982) Instinct and motivation as explanations for complex behavior. In *The Physiological Mechanisms of Motivation* (ed. D.W. Pfaff), Springer, New York, pp. 25–58

Everett, R.A., Ostfeld, R.S. and Davis, W.J. (1982) The behavioral hierarchy of the garden snail *Helix aspersa*. *Zeits. Tierpsychol.*, **59**, 109–26

Fantino, E. and Abarca, N. (1985) Choice, optimal foraging, and the delay-reduction hypothesis. *Beh. Brain Sci.*, **8**, 315–62

Feder, H.H. (1984) Hormones and sexual behaviour. *Ann. Rev. Psychol.*, **35**, 165–200

Fentress, J.C. (ed.) (1976) *Simpler Networks and Behavior*, Sinauer Assoc., Sunderland, MA

Fletcher, D.J.C. and Michener, C.D. (eds) (1987) *Kin Recognition in Animals*, Wiley, New York

Fraenkel, G.S. and Gunn, D.L. (1961) *The Orientation of Animals*, Dover, New York

Gallistel, C.R. (1980) *The Organization of Action: A New Synthesis*, Erlbaum, Hillsdale

Green, L., Rachlin, H. and Hanson, J. (1983) Matching and maximizing with concurrent ratio-interval schedules. *J. Exp. Anal. Beh.*, **40**, 217–24

Green, R.F. (1984) Stopping rules for optimal foragers. *Am. Nat.*, **123**, 30–40

Greenberg, N. and Maclean, P.D. (1978) *Behavior and Neurology of Lizards*, US Dept. Health Education and Welfare, Washington.

Greenberg, N., Chen, T. and Crews, D. (1984a) Social status, gonadal state, and the adrenal stress response in the lizard, *Anolis carolinensis*. *Horm. Beh.*, **18**, 1–11

Greenberg, N., Scott, M. and Crews, D. (1984b) Role of the amygdala in the reproductive and aggressive behavior of the lizard, *Anolis carolinensis*. *Physiol. Beh.*, **32**, 147–51

Griffin, D.R. (1984) *Animal Thinking*, Harvard University Press, Cambridge, MA

Guilford, T. and Dawkins, M.S. (1987) Search images not proven: A reappraisal of recent evidence. *Anim. Beh.*, **35**, 1838–45

Haccou, P., Dienske, M. and Meelis, E. (1983) Analysis of time inhomogeneity in Markov chains applied to mother–infant interactions of rhesus monkeys, *Anim. Beh.*, **31**, 927–45

Haccou, P. and Meelis, E. (1986) On the analysis of time-inhomogeneity in Markov chains: A refined test for abrupt behavioural changes. *Anim. Beh.*, **34**, 302–3

Halliday, T.R. (1975) An observational and experimental study of sexual behaviour in the smooth newt, *Triturus vulgaris* (Amphibia: Salamandridae). *Anim. Beh.*, **23**, 291–322

Halliday, T.R. (1976) The libidinous newt. An analysis of variation in the sexual behaviour of the male smooth newt, *Triturus vulgaris*. *Anim. Beh.*, **24**, 398–414

Halliday, T.R. (1977) The effect of experimental manipulation of breathing behaviour on the sexual behaviour of the smooth newt, *Triturus vulgaris*. *Anim. Beh.*, **25**, 39–45

Halliday, T.R. (1980) Motivational systems and interactions between activities. In *Analysis of Motivational Processes* (eds F.M. Toates and T.R. Halliday), Academic Press, London, pp. 205–20

Halperin, R. and Pfaff, D.W. (1982) Brain-stimulated reward and control of autonomic function: Are they related? In *The Physiological Mechanisms of Motivation* (ed. D.W. Pfaff), Springer, New York, pp.337–75

Hanlon, R.T., Hixon, R.F. and Hulet, W.H. (1983) Survival, growth, and behavior of the loliginid squids *Loligo plei, Loligo pealei,* and *Lolliguncula brevis* (Mollusca: Cephalopoda) in closed sea water systems. *Biol. Bull.,* **165**, 637–85

Harley, C.B. (1981) Learning the evolutionarily stable strategy. *J. Theor. Biol.,* **89**, 611–33

Heiligenberg, W. (1976) A probabilistic approach to the motivation of behavior. In *Simpler Networks and Behavior* (ed. J.C. Fentress), Sinauer Assoc., Sunderland, MA, pp. 301–13

Hinde, R.A. (1960) Energy models of motivation. *Soc. Exp. Biol. Symp.,* **14**, 199–213

Hinde, R.A. (1981) Animal signals: Ethological and games-theory approaches are not incompatible. *Anim. Beh.,* **29**, 535–42

Hinde, R.A. and Stevenson-Hinde, J.G. (eds) (1973) *Constraints on Learning,* Academic Press, London

Hinson, J.M. and Staddon, J.E.R. (1983) Matching, maximizing, and hill climbing. *J. Exp. Anal. Beh.,* **40**, 321–31

Hogan, J.A. (1965) An experimental study of conflict and fear: An analysis of behaviour of young chicks toward a mealworm. Part 1. The behavior of chicks which do not eat the mealworm. *Behaviour,* **25**, 45–97

Hogan, J.A. (1966) An experimental study of conflict and fear: An analysis of behaviour of young chicks toward a mealworm. Part 2. The behavior of chicks which eat the mealworm. *Behaviour,* **27**, 273–89

Hogan, J.A. (1971) The development of a hunger system in young chicks. *Behaviour,* **39**, 128–201

Hogan, J.A. (1977) Development of food recognition in young chicks: 4. Associative and nonassociative efforts of experience. *J. Comp. Physiol. Psychol.,* **91**, 839–50

Hogan, J.A. (1984) Pecking and feeding in chicks. *Learn. Motiv.,* **15**, 360–76

Hogan, J.A. and Roper, T.J. (1978) A comparison of the properties of different reinforcers. *Adv. Stud. Beh.,* **8**, 155–255

Holst, E.V. (1973) *The Behavioural Physiology of Animals and Man,* Methuen, London

Houston, A.I. (1982) Transitions and time sharing. *Anim. Beh.,* **30**, 615–25

Houston, A.I. (1983) Optimality theory and matching. *Beh. Anal. Lett.,* **3**, 1–15

Houston, A.I. (1986) The matching law applies to wagtails foraging in the wild. *J. Exp. Anal. Beh.,* **45**, 15–18

Houston, A.I. and McFarland, D.J. (1976) On the measurement of motivational variables. *Anim. Beh.*, **24**, 459–75

Hull, C.L. (1952) *A Behavior System*, Yale University Press, New Haven

Hulse, S.H., Fowler, H. and Honig, W.K. (1978) *Cognitive Processes in Animal Behavior*, Erlbaum, Hillsdale

Humphrey, N. (1983) *Consciousness Regained*, Oxford University Press., Oxford

Huntingford, F. and Giles, N. (1987) Individual variation in antipredator responses in the three-spined stickleback (*Gasterosteus aculeatus*). *Ethology*, **74**, 205–10

Huxley, J.S. (ed.) (1966) A discussion on ritualization of behaviour in animals and man. *Lond. Roy. Soc. Phil. Trans.*, **215B**, 247–526

Iersel, J.J.A.V. and Bol, A.C.A. (1958) Preening in two tern species: A study on displacement activities. *Behaviour*, **13**, 1–88

Ito, M., and Fantino, E. (1986) Choice, foraging, and reinforcer duration. *J. Exp. Anal. Beh.*, **46**, 93–103

James, W. (1902) *The Varieties of Religious Experience*, Longmans Green, New York

Johnson, D.F., Ackroff, K., Peters, J. and Collier, G.H. (1986) Changes in rats' meal patterns as a function of the caloric density of the diet. *Physiol. Beh.*, **36**, 929–36

Johnson, D.F. and Collier, G.H. (1987) Caloric regulation and patterns of food choice in a patchy environment: the value and cost of alternative foods. *Physiol. Beh.*, **39**, 351–9

Kacelnik, A. and Krebs, J.R. (1985) Learning to exploit patchily distributed food. In *Behavioural Ecology* (eds R.M. Sibly and R.H. Smith), Blackwell, Oxford, pp. 189–205

Kamil, A.C. and Yoerg, S.I. (1982) Learning and foraging behavior. In *Perspectives in Ethology. Vol.5* (eds P.P.G. Bateson and P.H. Klopfer), Plenum Press, New York, pp. 325–64

Kamil, A.C. and Roitblat, M.L. (1985) The ecology of foraging behavior: Implications for animal learning and memory. *Ann. Rev. Psychol.*, **36**, 141–69

Kamil, A.C., Krebs, J.R. and Pulliam, H.R. (eds) (1987) *Foraging Behavior*, Plenum Press, New York

Kandel, E.R. (1979) *Behavioral Biology of Aplysia*, Freeman, San Francisco

Kennedy, J.S. (1987) Animal motivation: The beginning of the end? In *Perspectives in Chemoreception and Behaviour* (eds R.F. Chapman, E.A. Bernays and J.G. Stoffolano Jr), Springer, New York, pp. 17–31

Kitcher, P. (1985) *Vaulting Ambition*, Massachusetts Institute of Technology, Cambridge, MA

Klir, G.J. (ed.) (1972) *Trends in General Systems Theory*, Wiley, New York

Knight, R.L. and Temple, S.A. (1986) Why does intensity of avian nest defence increase during the nesting cycle? *Auk*, **103**, 318–27

Kovac, M.P. and Davis, W.J. (1980a) Reciprocal inhibition between feeding and withdrawal behaviors in *Pleurobranchaea. J. Comp. Physiol.*, **139**, 77–86

Kovac, M.P. and Davis, W.J. (1980b) Neural mechanisms underlying behavioral choice in *Pleurobranchaea. J. Neurophysiol.*, **43**, 469–87

Krebs, J.R. (1983) From Skinner box to the field. *Nature*, **304**, 117

Krebs, J.R., Kacelnik, A. and Taylor, P. (1978) Test of optimal sampling by foraging great tits. *Nature*, **275**, 27–31

Krebs, J.R. and Dawkins, R. (1984) Animal signals: Mind-reading and manipulation. In *Behavioural Ecology* (eds J.R. Krebs and N.B. Davies), Sinauer Assoc., Sunderland, MA, pp. 380–402

Kuslansky, B., Weiss K.R. and Kupfermann, I. (1987) Mechanisms underlying satiation of feeding behavior of the mollusc *Aplysia. Beh. Neural Biol.*, **48**, 278–303

Lashley, K.S. (1938) Experimental analysis of instinctive behavior. *Psychol.Rev.*, **45**, 445–71

Lawrence, E.S. and Allen, J.A. (1983) On the term 'search image'. *Oikos*, **40**, 313–14

Lea, S. (1980) Supply as a factor in motivation. In *Analysis of Motivational Processes* (eds F.M. Toates and T.R. Halliday), Academic Press, London, pp. 153–77

Leonard, J.L. (1984) A top-down approach to the analysis of behavioural organization. *J. Theor. Biol.*, **107**, 457–70

Lester, N.P. (1976) Motivational implications of the temporal pattern of feeding in bluegill sunfish (*Lepomis macrochirus*). MSc thesis, Queen's University, Kingston, Canada

Lester, N.P. (1984a) The 'feed:feed' decision: How goldfish solve the patch depletion problem. *Behaviour*, **89**, 175–99

Lester, N.P. (1984b) The 'feed–drink' decision. *Behaviour*, **89**, 200–19

Lissak, K. and Molnar, P. (eds) (1982) *Motivation and the Neural and Neurohormonal Factors in Regulation of Behaviour*, Akademiai Kiado, Budapest

Lloyd Morgan, C. (1909) *An Introduction to Comparative Psychology, 2nd ed.*, Scott, London

Loeb, J. (1918) *Forced Movements, Tropism, and Animal Conduct*, Lippincott, Philadelphia

Lorenz, K. (1966) *On Aggression*, Harcourt, Brace & World, New York

Lorenz, K. (1971) *Studies in Human and Animal Behaviour, 2 vols.*, Methuen, London

MacCorquodale, K. and Meehl, P.E. (1948) On a distinction between hypothetical constructs and intervening variables. *Psychol. Rev.*, **55**,

95–107

Machlis, L. (1977) An analysis of the temporal patterning of pecking in chicks. *Behaviour*, **63**, 1–70

Machlis, L. (1980) Apomorphine: Effects on the timing and sequencing of pecking behavior in chicks. *Pharm. Bioch. Beh.*, **13**, 331–6

Machlis, L., Dodd P.W.D. and Fentress, J.C. (1985) The pooling fallacy: Problems arising when individuals contribute more than one observation to the data set. *Zeits. Tierpsychol.*, **68**, 201–14

Mackintosh, N.J. (1983) *Conditioning and Associative Learning*, Oxford University Press, Oxford

Magurran, A.E. (1986) Individual differences in fish behaviour. In *The Behaviour of Teleost Fishes* (ed T.J. Pitcher), Croom Helm, London, pp. 338–65

Mammen, D.L. and Nowicki, S. (1981) Individual differences and within-flock convergence in chickadee calls. *Beh. Ecol. Sociobiol.*, **9**, 179–86

Marcotte, B.M. and Browman, H.I. (1986) Foraging behaviour in fishes: Perspectives on variance. *Env. Biol. Fishes*, **16**, 25–33

Margoliash, D. and Konishi, M. (1985) Auditory representation of autogenous song in the song system of white-crowned sparrows. *US Proc. Nat. Acad. Sci.*, **82**, 5997–6000

Marler, P. and Sherman, V. (1983) Song structure without auditory feedback. Emendations of the auditory template hypothesis. *J. Neurosci.*, **3**, 517–31

Maynard Smith, J. (1982a) *Evolution and the Theory of Games*, Cambridge University Press, Cambridge

Maynard Smith, J. (1982b) Do animals convey information about their intentions? *J. Theor. Biol.*, **97**, 1–5

Mayr, E. (1982) *The Growth of Biological Thought*, Harvard University Press, Cambridge, MA

Mazur, J.E. (1981) Optimization theory fails to predict performance of pigeons in a two-response situation. *Science*, **214**, 823–5

McDougall, W. (1923) *Outline of Psychology*, Methuen, London

McFarland, D.J. (1970) Behavioral aspects of homeostasis. *Adv. Study Beh.*, **3**, 1–26

McFarland, D.J. (1971) *Feedback Mechanisms in Animal Behaviour*, Academic Press, London

McFarland, D.J. (ed.) (1974) *Motivational Control Systems Analysis*, Academic Press, London

McFarland, D.J. (1983) Time sharing: A reply to Houston (1982). *Anim. Beh.*, **31**, 307–8

McFarland, D.J. and Sibly, R.M. (1975) The behavioural final common path. *Lond. Roy. Soc. Phil. Trans.*, **270B**, 265–93

McFarland, D.J. and Houston, A.I. (1981) *Quantitative Ethology*, Pitman, Boston

McNair, J.N. (1980) A stochastic foraging model with predator training effects. 1. Functional response, switching and run lengths. *Theor. Pop. Biol.*, **17**, 141–66

McNair, J.N. (1981) A stochastic foraging model with predator training effects. 2. Optimal diets. *Theor. Pop. Biol.*, **19**, 147–62

McNamara, J. and Houston, A. (1980) The application of statistical decision theory to animal behaviour. *J. Theor. Biol.*, **85**, 673–90

McNamara, J.M. and Houston, A.I. (1985) Optimal foraging and learning. *J. Theor. Biol.*, **117**, 231–49

McNamara, J.M. and Houston, A.I. (1986) The common currency for behavioral decisions. *Am. Nat.*, **127**, 358–78

McNamara, J.M., and Houston, A.I. (1987) Partial preferences and foraging. *Anim. Beh.*, **35**, 1084–99

Metz, J.A.J., Dienske, H., DeJonge, G. and Putters, F.A. (1983) Continuous time Markov chains as models for animal behaviour. *Bull. Math. Biol.*, **45**, 643–58

Milinksi, M. (1984) Competitive resource sharing: An experimental test of a learning rule for ESSs. *Anim. Beh.*, **32**, 232–42

Miller, G.A., Galanter, E. and Pribram, K.H. (1960) *Plans and the Structure of Behavior*, Holt, New York

Miller, N.E. (1971) *Selected Papers, 2 vols.*, Aldine, Chicago

Morgan, C.T. (1943) *Physiological Psychology*, McGraw-Hill, New York

Morgane, P.J. (1975) Anatomical and neurobiochemical bases of the central nervous control of physiological regulations and behaviour. In *Neural Integration of Physiological Mechanisms and Behaviour* (eds G.J. Mogenson and F.R. Calarescu), University of Toronto Press, Toronto, pp. 24–67

Morris, D. (1957) 'Typical intensity' and its relation to the problem of ritualisation. *Behaviour*, **11**, 1–12

Morton, E.S. (1977) On the occurrence and significance of motivation-structural rules in some bird and mammal sounds. *Am. Nat.*, **111**, 855–69

Mowrer, O.H. (1960) *Learning Theory and Behavior*, Wiley, New York

Moynihan, M. (1982) Why is lying about intentions rare during some kinds of contests? *J. Theor. Biol.*, **97**, 9–12

Nagel, E. (1961) *The Structure of Science*, Harcourt, Brace & World, New York

Nelson, K. (1965) The temporal patterning of courtship behaviour in the glandulocaudine fishes (Ostariophysi, Characidae). *Behaviour*, **24**, 90–146

Noakes, D.L.G. (1986) When to feed: Decision making in sticklebacks, *Gasterosteus aculeatus. Env. Biol. Fishes*, **16**, 95–104

Norgren, R. and Grill, H. (1982) Brain-stem control of ingestive behavior. In *The Physiological Mechanisms of Motivation* (ed. D.W. Pfaff), Springer, New York, pp. 99–131

Nottebohm, F. (1984) Birdsong as a model in which to study brain processes related to learning. *Condor*, **86**, 227–36

Nottebohm, F., Nottebohm, M.E. and Crane, L. (1986) Developmental and seasonal changes in canary song and their relation to changes in the anatomy of song-control nuclei. *Beh. Neural Biol.*, **46**, 445–71

Nottebohm, F., Nottebohm, M.E., Crane L.A. and Wingfield, J.C. (1987) Seasonal changes in gonadal hormone levels of adult male canaries and their relation to song. *Beh. Neural Biol.*, **47**, 197–211

Ollason, J.G. (1980) Learning to forage – optimally? *Theor. Pop. Biol.*, **18**, 44–56

Ollason, J.G. (1983) Behavioural consequences of hunting by expectation: A simulation study of foraging tactics. *Theor. Pop. Biol.*, **23**, 323–46

O'Neill, R.V., DeAngelis D.L., Waide, J.B. and Allen, T.F.H. (1986) *A Hierarchical Concept of Ecosystems*, Princeton University Press, Princeton

Parker, G.A. (1984) Evolutionarily stable strategies. In *Behavioural Ecology* (eds J.R. Krebs and N.B. Davies), Sinauer Assoc., Sunderland, MA, pp. 30–61

Partridge, L. and Green, P. (1985) Intraspecific feeding specializations and population dynamics. In *Behavioural Ecology* (eds R.M. Sibly and R.H. Smith), Blackwell, Oxford, pp. 207–26

Patterson, F. (1978) Conversations with a gorilla. *Nat. Geogr.*, **154**, 438–65

Pavlov, I.P. (1927) *Conditioned Reflexes*, Oxford University Press, Oxford

Peeke, H.V.S. and Petrinovich L. (eds) (1984) *Habituation, Sensitization, and Behavior*, Academic Press, London

Pfaff, D.W. (ed.) (1982) *The Physiological Mechanisms of Motivation*, Springer, New York

Pierce, G.J. and Ollason, J.G. (1987) Eight reasons why optimal foraging theory is a complete waste of time. *Oikos*, **49**, 111–18

Plowright, C.M.S and Plowright, R.C. (1987) Oversampling by great tits? A critique of Krebs, Kacelnik, and Taylor's (1978) 'Test of optimal sampling by great tits'. *Can, J. Zool.*, **65**, 1282–3

Popper, K.R. (1962) *Conjectures and Refutations*, Routledge & Kegan Paul, London

Powley, T.L. (1977) The ventromedial hypothalamic syndrome, satiety, and a cephalic phase hypothesis. *Psychol. Rev.*, **84**, 89–126

Putters, F.A., Metz, J.A.J. and Kooijman, S.A.L.M. (1984) The identification of simple function of a Markov chain in a behavioural context: Barbs do it (almost) randomly. *Nieuw Arch. Wiskunde*, **2**, 110–23

Pyke, G.H. (1984) Optimal foraging theory: A critical review. *Ann. Rev. Ecol. Syst.*, **15**, 523–75

Rachlin, H. (1970) *Introduction to Modern Behaviorism*, Freeman, San Francisco

Regelmann, K. (1986) Learning to forage in a variable environment. *J. Theor. Biol.*, **120**, 321–9

Rhijn, J.G.V. (1980) Communication by agonistic displays: A discussion. *Behaviour*, **74**, 284–93

Rhijn, J.G.V. and Vodegel, R. (1980) Being honest about one's intentions: An evolutionarily stable strategy for animal conflicts. *J. Theor. Biol.*, **85**, 623–41

Richards, L.J. (1983) Hunger and the optimal diet. *Am. Nat.*, **122**, 326–34

Richelle, M., and Lejeune, H. (1980) *Time in Animal Behaviour*, Pergamon, Oxford

Richter, C.P. (1976) *The Psychobiology of Curt Richter*, York, Baltimore

Ristau, C.A. (1983) Language, cognition, and awareness in animals? *N.Y. Acad. Sci. Ann.*, **406**, 170–86

Ristau, C.A. (1986) Do animals think? In *Animal Intelligence* (eds R.J. Hoage and L. Goldman), Smithsonian, Washington, pp. 165–85

Rodger, R.S. and Rosebrugh, R.D. (1979) Computing a grammar for sequences of behavioural acts. *Anim. Beh.*, **27**, 737–49

Rohwer, S. (1978) Parent cannibalism of offspring and egg raiding as a courtship strategy. *Am. Nat.*, **112**, 429–40

Rohwer, S. and Ewald, P.W. (1981) The cost of dominance and advantage of subordination in a badge signalling system. *Evolution*, **35**, 441–54

Roitblat, H.L. (1987) *Introduction to Comparative Cognition*, Freeman, New York

Roper, T.J. (1980) Induced behaviour as evidence of nonspecific motivational effects. In *Analysis of Motivational Processes* (eds F.M. Toates and T.R. Halliday), Academic Press, London, pp. 221–42

Roper, T.J. (1984) Response of thirsty rats to absence of water: Frustration, disinhibition or compensation? *Anim. Beh.*, **32**, 1225–35

Roper, T.J. (1985) How plausible is post-inhibitory rebound as an account of displacement activity? A reply to Kennedy. *Anim. Beh.*, **33**, 1377–8

Rothstein, S.I. and Fleischer, R.C. (1987) Vocal dialects and their possible relation to honest signalling in the brown-headed cowbird. *Condor*, **89**, 1–23

Routtenberg, A. (1968) The two-arousal hypothesis: Reticular formation and limbic system. *Psychol. Rev.*, **75**, 51–80

Satinoff, E. and Teitelbaum, P. (eds) (1983) *Handbook of Behavioral Neurobiology, Vol. 6. Motivation*, Plenum Press, New York

Schleidt, W.M., Yakalis, G., Donnelly, M. and McGarry, J. (1984) A proposal for a standard ethogram, exemplified by an ethogram of the blue-breasted quail (*Coturnix chinensis*). *Zeits. Tierpsychol.*, **64**, 193–220

Schone, H. (1984) *Spatial Orientation*, Princeton University Press, Princeton

Sebeok, T.A. (ed) (1977) *How Animals Communicate*, Indiana University Press, Bloomington, IN

Sevenster, P. (1968) Motivation and learning in sticklebacks. In *The Central*

*Nervous System and Fish Behavior* (ed. D.J. Ingle), University of Chicago Press, Chicago, pp. 233–45

Sevenster, P. (1973) Incompatibility of response and reward. In *Constraints on Learning* (eds R.A. Hinde and J. Stevenson-Hinde), Academic Press, London, pp. 265–83

Sevenster, P. and Roosmalen, M.E.V. (1985) Cognition in sticklebacks: Some experiments on operant conditioning. *Behaviour*, **93**, 170–83

Sherrington, C.S. (1947) *The Integrative Action of the Nervous System*, Cambridge University Press, Cambridge

Shettleworth, S.J. (1985) Handling time and choice in pigeons. *J. Exp. Anal. Beh.*, **44**, 139–55

Sibly, R.M. (1980) The uses of mathematical models to describe behaviour sequences and to study their physiology and survival value. In *Analysis of Motivational Processes* (eds F.M. Toates and T.M. Halliday), Academic Press, London, pp. 245–71

Sibly, R.M. and McCleery, R.H. (1976) The dominance boundary method of determining motivational state. *Anim. Beh.*, **24**, 108–24

Sibly, R. and McFarland, D. (1976) On the fitness of behavior sequences. *Am. Nat.*, **110**, 601–17

Siegel, J.M. (1979). Behavioral functions of the reticular formation. *Brain Res. Rev.*, **1**, 69–105

Silver, R. (1978) The parental behavior of ring doves. *Am. Sci.*, **66**, 209–15

Skinner, B.F. (1974) *About Behaviorism*, Knopf, New York

Slater, P.J.B. (1981) Individual differences in animal behaviour. In *Perspectives in Ethology*. *Vol. 4* (eds P.P.G. Bateson and P.H. Klopfer), Plenum Press, New York, pp. 35–49

Slater, P.J.B. and Lester, N.P. (1982) Minimising errors in splitting behaviour into bouts. *Behaviour*, **79**, 153–61

Smith, K. (1984) 'Drive': In defense of a concept. *Behaviorism*, **12**, 71–114

Smith, W.J. (1977) *The Behavior of Communicating*, Harvard University Press, Cambridge, MA

Snyderman, M. (1983) Optimal prey selection: The effects of deprivation. *Beh. Anal. Letts.*, **3**, 359–69

Solomon, R.L. (1982) The opponent processes in acquired motivation. In *The Physiological Mechanisms of Motivation* (ed. D.W. Pfaff), Springer, New York, pp. 321–36

Staddon, J.E.R. (1980) *Limits to Action*, Academic Press, New York

Staddon, J.E.R. and Hinson, J.M. (1983) Optimization: A result or a mechanism? *Science*, **221**, 976–7

Stearns, S.C. and Hempel, P.S. (1987) Evolutionary insights should not be wasted. *Oikos*, **49**, 118–25

Stellar, J.R. and Stellar, E. (1985) *The Neurobiology of Motivation and Reward*, Springer, New York

Stephens, D.W. and Krebs, J.R. (1986) *Foraging Theory*, Princeton University Press, Princeton.

Stolerman, I.P. (1985) Motivational effects of opioids. Evidence on the role of endorphins in mediating reward or aversion. *Pharm. Bioch. Beh.*, **23**, 877–81

Thompson, T. and Lubinski, D. (1986) Units of analysis and kinetic structure of behavioural repertoires. *J. Exp. Anal. Beh.*, **46**, 219–42

Thorndike, E.L. (1911) *Animal Intelligence*, Macmillan, New York

Tinbergen, N. (1951) *The Study of Instinct*, Oxford University Press, Oxford

Tinbergen, N. (1952) 'Derived' activities; their causation, biological significance, origin, and emancipation during evolution. *Quart. J. Biol.*, **27**, 1–32

Tinbergen, N. (1963) On aims and methods of ethology. *Zeits. Tierpsychol.*, **20**, 410–33

Toates, F.M. (1980) *Animal Behaviour – A Systems Approach*, Wiley, Chichester

Toates, F.M. (1986) *Motivational Systems*, Cambridge University Press, Cambridge

Toates, F.M. and O'Rourke, C. (1978) Computer simulation of male rat sexual behaviour. *Med. Biol. Engin. Comput.*, **16**, 98–104

Toates, F.M. and Halliday, T.M. (eds) (1980) *Analysis of Motivational Processes*, Academic Press, London

Tolman, E.C. (1951) *Collected Papers in Psychology*, University of California Press, Berkeley

Turner, G.F. and Huntingford, F.A. (1986) A problems for game theory analysis: Assessment of intention in male mouthbrooder contests. *Anim. Beh.*, **34**, 961–70

Vanderwolf, C.H. (1983) The role of the cerebral cortex and ascending activating systems in the control of behavior. In *Handbook of Behavioral Neurobiology. Vol. 6 Motivation* (eds E. Satinoff and P. Teitelbaum), Plenum Press, New York, pp. 67–104

Veen, J. (1987) Ambivalence in the structure of display vocalizations of gulls and terns: New evidence in favour of Tinbergen's conflict hypothesis? *Behaviour*, **100**, 33–49

Wallace, B. (1973) Misinformation, fitness, and selection. *Am. Nat.*, **107**, 1–7

Walter, W.G. (1953) *The Living Brain*, Norton, New York

Watson, J.B. (1930) *Behaviorism*, University of Chicago Press, Chicago

Weichel, K., Schwager, G., Heid, P., Guttinger, H.R. and Pesch, A. (1986) Sex differences in plasma steroid concentrations and singing behaviour during ontogeny in canaries (*Serinus canaria*). *Ethology*, **73**, 281–94

Weingarten, H.P. and Powley, T.L. (1981) Pavlovian conditioning of the cephalic phase of gastric acid secretion in the cat. *Physiol. Beh.*, **27**, 217–21

Wells, M.J. (1978) *Octopus*, Chapman and Hall, London

Whalen, R.E. and Simon, N.G. (1984) Biological motivation. *Ann. Rev. Psychol.*, **35**, 257–76

Wiener, N. (1961) *Cybernetics or Control and Communication in the Animal and Machine. 2nd ed.*, Wiley, New York

Williams, B.A. (1986) On the role of theory in behaviour analysis. *Behaviorism*, **14**, 111–24

Williams, H. (1985) Sexual dimorphism of auditory activity in the zebra finch song system. *Beh. Neur. Biol.*, **44**, 470–84

Wilson, E.O. (1975) *Sociobiology*, Harvard University Press, Cambridge, MA

Wilson, J.Q. and Herrnstein, R.J. (1985) *Crimes and Human Nature*, Simon & Schuster, New York

Wilz, K.J. (1970a) Causal and functional analysis of dorsal pricking and nest activity in the courtship of the three-spined stickleback *Gasterosteus aculeatus*. *Anim. Beh.*, **18**, 115–24

Wilz, K.J. (1970b) The disinhibition interpretation of the displacement activities during courtship in the three-spined stickleback, *Gasterosteus aculeatus*. *Anim. Beh.*, **18**, 682–7

Woodside, B. and Milleline, L. (1987) Self-selection of calcium during pregnancy and lactation in rats. *Physiol. Beh.*, **39**, 291–6

Woodworth, R.S. (1918) *Dynamic Psychology*, Columbia University Press, New York

Wyers, E.J. (1985) Cognitive behavior and sticklebacks. *Behaviour*, **95**, 1–10

Yaniv, Y. and Golani, I. (1987) Superiority and inferiority: A morphological analysis of free and stimulus bound behaviour in honey badger (*Mellivora capensis*) interactions. *Ethology*, **74**, 89–116

Young, P.T. (1961) *Motivation and Emotion*, Wiley, New York

Zahavi, A. (1980) Ritualization and the evolution of movement signals. *Behaviour*, **72**, 77–81

Zeigler, H.P. (1964) Displacement activity and motivational theory. A case study in the history of ethology. *Psychol. Bull.*, **61**, 362–76

Zucker, I. (1983) Motivation, biological clocks, and temporal acquisition of behaviour. In *Handbook of Behavioral Neurobiology. Vol. 6 Motivation* (eds E. Satinoff and P. Teitelbaum), Plenum Press, New York, pp. 3–21

# Author index

Achinstein, P. 63
Ackroff, K. 71
Armstrong, E.A. 48
Aschoff, J. 20
Ashby, W.R. 12
Atkinson, J.W. 5
August, P.V. 125

Baerends, G.P. 38–42, 48, 123
Baker, M.C. 126
Bambridge, R. 82
Barlow, G.W. 13, 131
Barnes, W.J.P. 12, 56
Bateson, P.P.G. 26
Bell, W.J. 8, 15
Bernard, C. 77
Bertalanffy, L.V. 12
Berthold, A.A. 92
Berthoud, H.R. 90
Beukema, J. 119
Bindra, D. 5, 33
Boden, M. 64
Bolles, R.C. 5
Bonsall, R.W. 21
Bookstaber, R. 112
Booth, D.A. 91
Bowdan, E. 72
Brenner, S. 71
Bridgman, P.W. 63
Broom, D.M. 22, 26–27
Brown, J.A. 14
Brown, J.S. 5
Brown, R.E. 52
Bunge, M. 69

Cabanac, M. 64
Calow, P. 12
Campbell, B.G. 126

Campfield, L.A. 90–1
Cannon, W.B. 29, 77
Caryl, P.G. 128, 129
Cassidy, J. 7
Churchland, P.S. 69
Clark, A.B. 116
Cofer, C.N. 5
Cohen, J. 63
Cohen, S. 53
Colgan, P.W. 17, 38, 43, 44, 52,
    57–9, 118, 119
Collier, G.H. 19, 101, 120–2
Cowie, R.J. 106–7
Cox, J.E. 89–90
Crawford, L.I. 114
Crespi, L.P. 34, 63
Crews, D. 96
Croll, R.P. 76

Darwin, C.R. 124
Davis, W.J. 74–6, 98, 99
Dawkins, M. 65
Dawkins, R. 16, 17, 52
DeCarlo, L.T. 107–10
Dennett, D.C. 67, 127
Dethier, V.G. 71–3, 98–9
Diamond, C. 127
Dickenson, A. 33
Dill, L.M. 118–20
Douglas, J.M. 17–18
Drummond, H. 13
Dubos, R. 133
Duggan, J.P. 91

Elliott, J.P. 120
Epstein, A.N. 7, 8
Everett, R.A. 75

Fantino, E. 113
Feder, H.H. 92
Fentress, J.C. 71
Fletcher, D.J.C. 118
Fraenkel, G.S. 8

Gallistel, C.R. 70
Green, L. 110
Green, R.F. 112
Greenberg, N. 96–8
Griffin, D.R. 65
Guilford, T. 118

Haccou, P. 60–2
Halliday, T.R. 18, 23, 56, 123
Halperin, R. 86
Hanlon, R.T. 39
Harley, C.B. 112–13
Heiligenberg, W. 44, 45, 56
Hinde, R.A. 2, 8, 35, 128
Hinson, J.M. 113
Hogan, J.A. 26–32, 34–5, 78–80,
   101, 116
Holst, E.V. 70
Houston, A.I. 43, 53, 107, 112
Hull, C.L. 5, 8
Hulse, S.H. 63
Humphrey, N. 66–7
Huntingford, F. 117
Huxley, J.S. 125

Iersel, J.J.A.V. 48
Ito, M. 113

James, W. 29, 66
Johnson, D.F. 121

Kacelnik, A. 113, 118
Kamil, A.C. 105, 106, 110, 118
Kandel, E.R. 73–4
Kennedy, J.S. 99
Kitcher, P. 4
Klir, G.J. 12
Knight, R.L. 35
Kovac, M.P. 74–5
Krebs, J.R. 109, 110, 126
Kuslansky, B. 76

Lashley, K.S. 71
Lawrence, E.S. 118
Lea, S. 105
Leonard, J.L. 24
Lester, N.P. 55, 104, 113
Lissak, K. 70
Lloyd Morgan, C. 67
Loeb, J. 8, 12
Lorenz, K. 2, 4, 8, 15, 46, 124

MacCorquodale, K. 3, 63
Machlis, L. 20, 21, 24, 25, 115
Mackintosh, N.J. 33
Magurran, A.E. 117
Mammen, D.L. 118
Marcotte, B.M. 65, 117
Margoliash, D. 92
Marler, P. 65
Maynard Smith, J. 117, 128, 129
Mayr, E. 115
Mazur, J.E. 108–9
McDougall, W. 5
McFarland, D.J. 8, 12, 47, 52–6,
   102–4
McNair, J.N. 112
McNamara, J.M. 101, 112, 113,
   115
Metz, J.A.J. 60
Milinksi, M. 113
Miller, G.A. 64
Miller, N.E. 47
Morgan, C.T. 73
Morgane, P.J. 99
Morris, D. 47
Morton, E.S. 124
Mowrer, O.H. 64
Moynihan, M. 128

Nagel, E. 69
Nelson, K. 56
Noakes, D.L.G. 54–5
Norgren, R. 87
Nottebohm, F. 92–5

Ollason, J.G. 112
O'Neill, R.V. 134

Parker, G.A. 128

Partridge, L. 116–17
Patterson, F. 127
Pavlov, I.P. 8, 32
Peeke, H.V.S. 35
Pfaff, D.W. 70
Pierce, G.J. 106
Plowright, C.M.S. 110
Popper, K.R. 66
Powley, T.L. 87–9
Putters, F.A. 20, 61
Pyke, G.H. 106

Rachlin, H. 65
Regelmann, K. 111
Rhijn, J.G.V. 128
Richards, L.J. 120
Richelle, M. 20
Richter, C.P. 77
Ristau, C.A. 66, 131
Rodger, R.S. 17
Rohwer, S. 126, 127
Roitblat, H.L. 63
Roper, T.J. 49
Rothstein, S.I. 127
Routtenberg, A. 83–4

Satinoff, E. 70
Schleidt, W.M. 13
Schone, H. 8, 12, 56
Sebeok, T.A. 123
Sevenster, P. 35–7, 65
Sherrington, C.S. 8, 70
Shettleworth, S.J. 110
Sibly, R.M. 56, 102–4
Siegel, J.M. 81
Silver, R. 93–4
Skinner, B.F. 5
Slater, P.J.B. 20, 115, 117
Smith, K. 8
Smith, W.J. 127
Snyderman, M. 120

Solomon, R.L. 5, 64
Staddon, J.E.R. 105, 109
Stellar, J.R. 76, 86
Stearns, S.C. 106
Stephens, D.W. 106
Stolerman, I.P. 70

Thompson, T. 16
Thorndike, E.L. 5, 32
Tinbergen, N. 1, 2, 4, 7, 8, 13, 17,
    25, 26, 51, 125
Toates, F.M. 5, 12, 38, 56, 59–60
Tolman, E.C. 5
Turner, G.F. 129–31

Vanderwolf, C.H. 84–6
Veen, J. 123–5

Wallace, B. 128
Walter, W.G. 12
Watson, J.B. 63
Weichel, K. 94
Weingarten, H.P. 88
Wells, M.J. 39
Whalen, R.E. 70
Wiener, N. 12
Williams, B.A. 63
Williams, H. 95–6
Wilson, E.O. 11
Wilson, J.Q. 4
Wilz, K.J. 50
Woodside, B. 71
Woodworth, R.S. 8
Wyers, E.J. 65

Yaniv, Y. 13
Young, P.T. 5, 29, 48, 64

Zahavi, A. 118, 125
Zeigler, H.P. 48
Zucker, I. 20

# Subject index

Acquired motivation  32
Acquisition  33
Action specific energy  45
Adaptationism  65, 101
Allochthonous causation  49
Alternating act  47
Ambivalent act  47
Appetitive stimulus  4, 37
Associations  3, 5
Authochthonous causation  49

Behavioural final common path  47
Behaviourism  63

Catastrophe theory  57
Causal factor space  102
Causal system  6, 17
Causation  2, 80, 133
Central excitatory state  71
Central motive state  73
Chaining  52
Choices  19
Classical conditioning  32
Classical ethology  2
Cognitive framework  6, 11, 63
Communication  123
Competition  52
Concurrent variables  42
Concurrent schedule  107
Conditioning  32
Conflict  40, 47, 123
Constraints  32
Consummatory stimulus  4, 37
Corollary discharge  77
Correlations  17
Crystallization  94
Currency  101
Cybernetics  6, 12, 59, 64

Delay-reduction hypothesis  113
Diet breadth  120
Disinhibition hypothesis  48
Displacement activities  48
Dominance  52
Dominance boundary  102
Drive  6, 133
Drugs  70, 83

Economy, closed or open  115
Efference copy  77
Emancipation  125
Emergentism  67
Emotions  6, 29, 48, 64
Essentialism  115
Ethogram  13
Evolutionarily stable strategy  117,
    128
Excitation  48
Extinction  33

Feedback and feedforward  12, 55,
    133
Fixed action pattern (FAP)  15
Foraging  118
Frustration  48
Function  2, 80, 133

Give-up time  106
Goals  1, 2, 64
Grammars  17

Habituation  35, 73
Hardware approach to
    motivation  99
Hedonism  5, 33, 64
Hierarchy  16, 52, 75, 134
Hill-climbing  113

Homeostasis 77
Honesty of signals 126
Hormic theory 5
Hormones 92
Hunger 86, 118
Hypothetical construct 63

Incentive mechanisms 34, 37, 64, 133
Individual differences 115
Inhibition 48
Instinct 6, 134
Intention 123
Intention movements 47
Interactions of motivational systems 46
Interruption experiments 53
Interruptive activities 48
Intervening variable 63

Law of Heterogeneous Summation 38
Learning 32
Lesions 69, 86
Local contrast 144
Local enhancement 38
Logical positivism 63
Log-survivorship plot 19

Markov probability model 17, 60
Matching 105
Maximizing 105
Memory window 114
Metabolism 90
Modal action pattern (MAP) 13, 115
Motivation 3
Motivation-structural rules 124

Ontogeny 25
Operant conditioning 32
Operationalism 63
Opponent processes 5
Optimization 101, 102, 105

Patchy environment 120
Patterning 3
Post-inhibitory rebound 50

Preferences 19, 21
Pre-post state histogram 17
Priming 26, 33, 133
Principle of Antithesis 124
Proximate causation 2, 101
Psychophysics 56

Quantification 6, 11, 17, 41, 56, 134

Recognition 118
Redirected response 48
Reductionism 69
Reflex 8, 87
Reinforcement 33, 64
Relative Payoff Sum Rule 112–13
Releaser 26, 37, 133
Reward 33
Rhythms 20
Risk in foraging 111, 120
Ritualization 125
Rules of thumb 111

Sampling 110
Satisficing 112
Search images 118
Sensitization 35, 73
Sequences 16
Sham feeding 89
Signalling 123
Sleep 80
Social facilitation 38
Sociobiology 6
Software approach to motivation 99
Specific hungers 71
Spontaneous behaviour 46
State space model 102
Stereotypy 16, 94
Stimulus 3, 37, 133
Structural changes 3
Summation of causal factors 37
Switching 52
Symbolic inferiorism 40
Systems theory 6, 11

Taxonomy 13
Temporal patterning 19
Thwarting 47

Time-sharing 52, 103
Titration 42
Top-down approach 24
Trajectory 102
Transitivity 19
Types, theory of 66
Typical intensity 47

Typology 115

Ultimate causation 2, 101

Vacuum activities 44

Zeitgeber 21